THE
MILLION
DEATH
QUAKE

THE
MILLION
DEATH
QUAKE

THE SCIENCE OF PREDICTING EARTH'S DEADLIEST NATURAL DISASTER

ROGER MUSSON

palgrave
macmillan

First published in 2012 by PALGRAVE MACMILLAN® in the U.S.—a division of St. Martin's Press LLC, 175 Fifth Avenue, New York, NY 10010.

Where this book is distributed in the UK, Europe and the rest of the world, this is by Palgrave Macmillan, a division of Macmillan Publishers Limited, registered in England, company number 785998, of Houndmills, Basingstoke, Hampshire RG21 6XS.

Palgrave Macmillan is the global academic imprint of the above companies and has companies and representatives throughout the world.

Palgrave® and Macmillan® are registered trademarks in the United States, the United Kingdom, Europe and other countries.

ISBN: 978-0-230-11941-3

Library of Congress Cataloging-in-Publication Data

Musson, Roger.
    The million death quake : the science of predicting Earth's deadliest natural disaster / Roger Musson.
        p.   cm.
    Includes index.
    ISBN 978-0-230-11941-3 (hardback)
    1. Earthquake prediction. 2. Earthquakes. I. Title.
QE538.8.M87   2012
551.22—dc23

                                                          2012011137

A catalogue record of the book is available from the British Library.

Design by Letra Libre

First edition: October 2012

10   9   8   7   6   5   4   3   2   1

Printed in the United States of America.

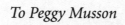

To Peggy Musson

CONTENTS

ACKNOWLEDGMENTS

IT HAS BEEN MY PRIVILEGE AND PLEASURE, OVER THE course of my career, to meet many of the outstanding seismologists of my generation, from whom I have absorbed much. In writing this book, the problem was not so much what to put in as what to leave out. I'm grateful to all those who have shared ideas with me over the years.

The idea for this book, and the first rough outline, actually came to me some years ago, as I was walking one evening through the streets of Bucharest on my way back to the hotel where I was staying. At that time it progressed no further than some notes entered on my laptop when I returned to my room. I owe a great debt to my editor, Laura Lancaster, for enabling me to turn those first ideas into a viable book. Editors have a tough time; they get blamed for the faults in a book, but seldom do critics give them credit when things go right.

Writing this book would also have been difficult without the care and constant support of my wife, Hazel, for which I am deeply grateful.

It was always a disappointment to my mother that no one ever dedicated a book to her. I'm sorry I could not have repaired the omission in her lifetime, but I dedicate this book to her memory.

A NOTE ABOUT UNITS

IT IS A UNIVERSAL RULE IN SCIENTIFIC PUBLICATIONS THAT all measurements are expressed in decimal units, and I have generally adhered to this. Particularly, one cannot write, "such-and-such was reported to be twenty miles deep" when the original report would inevitably have given the distance in kilometers only. However, sometimes it's easier to visualize things in imperial units, and occasionally older documents cited really did use miles.

HOTSPOTS AND ROGUE EARTHQUAKES

The shaded areas of this map show the major earthquake zones of the world, with the exception of those that are oceanic. This should be treated as a sketch map only; the zones are not precisely defined and the density of shading is largely impressionistic. Stars show the location of large earthquakes outside the shaded zones, with the exception of some offshore earthquakes. Some of these locations have had repeated earthquakes (sometimes only detectable from the geological record) while others seem to be "one-off."

PART 1
PROBLEMS

1
SCREAMING CITIES

IMAGINE A CITY SCREAMING.

A large city with a population of millions, perhaps about the size of Chicago or Madrid, and everywhere people are screaming. Not just on one block, as might happen if there was an accident or an explosion. Block after block, street after street—in the downtown, by the harbor, in the suburbs, everywhere people are screaming. The traffic has stopped, people are getting out of their cars; many people are running, others are milling around in a state of shock; still others are just standing, clutching their faces, and screaming. Many do nothing at all, because they are dead. Everywhere it is the same.

Five minutes ago it was all different. Just five minutes ago it was a normal afternoon. The streets were busy with people going about their day-to-day business; the market stalls were doing their usual trade. Some people were already making their way home, carrying shopping baskets of vegetables for dinner, their minds on plans for the evening. Just another day like any other, and people's minds were on the usual things: "Those oranges look good"; "I ought to get a new shirt"; "I must fix the roof tomorrow." All the different thoughts, conversations, and activities that make up the life of a busy, bustling city.

How quickly it all changed. There was no warning. Suddenly there was a terrific sound like an express train rushing down the center of the street, and everything started rocking. Then, from all around, came the sound of breaking glass and then the baleful rumble of collapsing masonry. And then screaming.

And dust. As the buildings caved in and turned to rubble, clouds of dust rose into the air, a horrible yellowy-gray dust of powdered concrete and cement. And it was everywhere. Buildings were collapsing across the city, each sending up its own cloud of dust. A woman standing on her balcony overlooking the city saw the clouds merge and the whole city disappear in one huge rising cloud of dust. No streets, no buildings, no landmarks, just dust that obscured everything. But she could hear the roar of collapsing buildings and the shrieks and screams. The woman was already on the phone to a distant friend when the disaster struck. She cried into the receiver, "It's the end of the world!"

This was Port-au-Prince, Haiti, on January 12, 2010. The local time was seven minutes to five in the afternoon.

AT THE TIME OF THE HAITI EARTHQUAKE I WAS SITTING IN my study at home in Edinburgh. The time was five hours later than in Haiti. At about 10:15 P.M., as the first waves from the earthquake were beginning to reach the seismometers of the UK monitoring network, the phone rang. A journalist from the BBC had already heard about the earthquake and wanted to know what I could tell him about it.

A hundred years ago, if an earthquake happened in a distant place, the seismologists usually were the first to know, as the first seismic waves arrived at their instruments. Nowadays communications are vastly different, and radio waves travel much faster than sound, so when an earthquake occurs, the news spreads around the globe almost at once. Journalists can often be first with the news.

Thanks to the connectedness of the Internet and the great work done in recent years by the seismological community (especially those with the US Geological Survey) in exchanging data and linking systems, it was not difficult to bring up the first results on my computer screen.

These were being compiled automatically from data gathered by those monitoring stations closer to Haiti that had already picked up signals and started computing earthquake parameters.

From what I could see, the news was all bad. I had worked some years previously on a project in the Dominican Republic, next door to Haiti, and I had some familiarity with the complex geological structure of the region and how it related to earthquakes. In particular, I knew about an especially dangerous fault that runs along the southern peninsula of Haiti and into the Dominican Republic. Blessed with the rather baroque name of the Enriquillo–Plantain Garden Fault, it is a structure similar to the much-better-known San Andreas Fault in California. And similarly deadly.

And this earthquake was right on it. And right next to the capital city of Haiti—a city whose population was densely packed into badly built houses that straggled up into the surrounding shantytowns. A city cursed with poverty and already weakened by recent hurricanes and floods. A city as vulnerable to earthquakes as ever a city could be. And here was a major earthquake scoring practically a direct hit.

What I was thinking was, "Population of most-affected area—about three million; likely fatality rate—maybe 5 to 10 percent; likely death toll—150,000? 300,000?" What I said was, "I'm afraid this is going to be very bad."

But how bad? The true extent of the devastation after an earthquake is often slow to emerge. Usually it follows a characteristic pattern. The first news reports of the disaster usually say something like "At least twelve people are confirmed to have been killed," which probably means that the reporter was able to count twelve bodies before filing the story. The next report says, "At least two hundred people are now known to have died." A day later the figure is "more than a thousand," and so it goes. Eventually the official death toll for the Haiti earthquake reached 220,000, a figure cited frequently in accounts of the earthquake published during the following year. Then, finally, twelve months later a revised figure was released—316,000 dead. Officially the second-deadliest earthquake in human history. The figure is disputed, and often one never really knows the truth. In the first days after an earthquake, the prime consideration is rescuing those trapped under the wreckage and saving as many lives as possible. All effort must be directed toward helping the living; the dead are beyond help. So making an accurate count of the dead is not a high priority. Then, for health reasons, it's often necessary to clear away corpses as quickly as possible, even if that means mass burial. So in the chaos following a major earthquake disaster, establishing an accurate body count can be nearly impossible.

But even if establishing an accurate ranking of the ten deadliest earthquakes is not possible, Haiti is clearly up there with the worst—somewhere in the top ten, if not actually second.

A TALE OF TWO EARTHQUAKES

Just over a month later, on February 27, the news once again was full of stories about an earthquake, this time in Chile. A truly massive earthquake—hundreds of times larger than the Haiti earthquake—struck off the coast of central Chile. The earth's crust was violently displaced by as much as ten meters as rocks were ripped apart for seven hundred kilometers along the Chilean coast, devastation that took three minutes from start to finish.

This huge earthquake proved a different story from the Haiti quake. Once again the death toll started small, then gradually rose and rose—and then came down again. From an estimate of 802, it declined to 521 as some people who had been marked as missing and presumed dead turned out to be very much alive, having taken refuge with relatives outside the disaster zone.

Chile is very different from Haiti. It may not be the richest country in the world, but it is not mired in poverty like Haiti. And, crucially, everyone in Chile knows the country has an earthquake problem.

Earthquakes are common there. As a result buildings are routinely built to withstand earthquakes. Even if Haiti had had the resources to make buildings safer, no one in Port-au-Prince had ever felt an earthquake there before, and people had no idea they were at risk.

These two quakes from early 2010 make an instructive comparison. People sometimes imagine that the larger an earthquake is, the worse it must be. In truth, many factors determine an earthquake's impact.

The Haiti earthquake was something like the worst case imaginable. It was strong, and it was close to a major city. The Chilean quake might have been larger, but it didn't deliver the same concentrated punch to one major city. And, most important, the Haitians were far more vulnerable. They were not prepared.

An Australian geologist of my acquaintance, Ted Brennan, devised his own personal scale of earthquake disaster potential, rating cities from one to ten. This wasn't based on frequency of earthquakes—almost the reverse. He would give a low rating to cities that have often been exposed to earthquakes. The high ratings he awarded to cities that had been hit in the past but well beyond living memory, places where people were complacent in the belief that earthquakes were not their problem. The last time Port-au-Prince had been destroyed by an earthquake was 1770, when it was a lot smaller. Who in modern Haiti knew anything about that?

Chile might be a far more earthquake-prone country, but it was prepared, and the preparations by and large did their work.

Haiti 2010 was one of the deadliest earthquakes on record, and Chile 2010 was one of the largest. Oddly enough, if one tries to list the largest quakes ever and the deadliest quakes ever, few appear on both lists. But both Haiti's and Chile's were quakes for the record books, and the Chile quake occurred with the harrowing accounts from Haiti still fresh in everyone's mind. It caused a lot of people to take notice. Two disasters so close together. Could it be that earthquakes are becoming more common?

On some Internet bulletin boards a conspiracy theory started to develop—that earthquakes are becoming more frequent, but seismologists have plotted to conceal the alarming truth. At the same time a rumor was circulating that climate scientists were plotting to exaggerate the risks of climate change in order to get more research dollars. So why, exactly, would seismologists be trying to conceal a growing peril instead of looking for more funding for seismology? However, the question of whether earthquakes are becoming more frequent is common and needs an answer. In fact, it is asked so often that most seismologists have a stock reply. The odd thing is that when you tell people the answer is no, they often seem disappointed, as though more earthquakes would be a good thing.

There are four ways of looking at the question. The first is to look simply at the facts. We have the statistics. We have catalogs that go back several hundred years in some parts of the globe. It is true that more earthquakes are recorded today than in previous decades, because there are more seismometers, but these are all small earthquakes. Once you compare like with like, you find that the rate at which large earthquakes occur is stable throughout history.

Most people have no idea how many earthquakes occur every day, because they aren't news. An earthquake in the middle of the Atlantic that has no human impact is not going to be reported in the general media. The general public hears about only those earthquakes that happen in populated areas and, even then, usually only if someone is killed. So the frequency with which earthquakes make the news is different from the frequency with which they occur and that seismologists see.

In 1976 high-profile earthquake disasters occurred in Guatemala, China, Turkey, the Philippines, and Italy. Whether earthquakes were becoming more frequent was widely debated in the press. In fact at the end of the year it turned out that 1976 was not a particularly seismic year—it just had more newsworthy quakes than usual.

And this brings me to the third issue: people have short memories. When people asked about earthquake frequency in 2010, they probably were not thinking further back than 2009. Certainly no one was comparing recent earthquake activity with the earthquakes of 1976. I'm waiting for someone to ask why we are having so few earthquakes in one of those periods when there is a lull, but somehow that question never comes. People notice the peaks in activity but never the troughs.

Last, where could the energy for an increase in earthquakes come from? A major earthquake can release energy equivalent to hundreds of nuclear weapons. Ultimately, the energy source is heat in the planet's interior. A significant increase in the number of earthquakes would require a sudden increase in temperature deep inside the earth, and this doesn't seem very likely. Hence one can expect earthquakes to occur at a fairly constant rate; Chapter 3 will be the place to go into this in more detail.

Haiti again provides an instructive example. In human terms it was a calamity. In geological terms it was unexceptional. On average, earthquakes the same size as the Haiti quake occur ten or twelve times a year around the globe. As I write these words, I have just heard news of the latest—in Siberia. Estimated impact: nil. The conspiracy theorists probably won't even hear about that one.

But the unchanged frequency of earthquakes is not the full story. No one worries much about an earthquake in the middle of the Atlantic because nothing is there to be damaged. Earthquakes are a problem when they strike cities. And cities are getting larger. As the world's urban population grows, the danger that an earthquake will be a disaster rather than a statistic grows too.

THE DESERT OF DEATH

Pick up any volume of a third-rate fantasy trilogy and you should find, somewhere at the front, a map of the fantasy land in which the story is set. A picture of an erupting volcano (labeled "Mount Doom") usually appears somewhere in a corner, and somewhere on the map you will probably find a desert called "The Empty Desert" or perhaps "Desert of Death"—or you might even find both. Generally they will be marked with a little picture of a skeleton to emphasize that these are not good places to be.

As it happens, both are real place names. They sound rather more exotic in Persian, though: Dasht-i-Lut and Dasht-i-Margo, desert of emptiness and desert of death. They sit in eastern Iran, with the Dasht-i-Margo extending well into Afghanistan, and, as the names suggest, they are two of the most inhospitable places on the planet. Viewed from space (or Google Earth) they show up as vast bruises of purple, brown, and gray. On the ground they are endless expanses of sand and gravel, broken by pillars of rock that the wind has carved into surreal shapes. The Dasht-i-Lut has the distinction of being the hottest place on Earth; surface temperatures can reach as high as 70 degrees Celsius.

And yet sandwiched between these two terrible deserts is a village, the little settlement of Sefidabeh, where a few hundred people manage to sustain themselves from crops grown with the aid of irrigation from several springs of freshwater. In February 1994 a strong earthquake hit the region. If it had occurred anywhere in the middle of those empty deserts, it would hardly have merited attention. But it struck precisely under Sefidabeh, flattening the three hundred or so houses. Fortunately for the villagers, the earthquake hit at midmorning, when everyone was working outdoors, so few people were killed.[1]

That an earthquake should score a bull's-eye on the one little village in the middle of nowhere seems astonishing bad luck. But was it really bad luck, or was there something more to it?

The answer lies in the one factor that allowed people to start a village in such a dismal desert region: water. The only place anyone could live was next to those springs. And the springs were there because of a fault in the rocks, a crack through which water could bubble up and reach the surface. But an active fault can still move and cause earthquakes, and that was what happened at Sefidabeh. Ironically the one thing that allowed the village to exist also guaranteed its eventual destruction.

So the forces that cause earthquakes can also create desirable places to live. Volcanoes present a similar problem: volcanic soil is highly fertile and great for growing crops. This attracts people to exactly the places where they will be hit by the next major eruption. When the eruption is over, people move back to the good soil, confident that the volcano will probably not erupt again in their lifetime. Eventually the next eruption overwhelms their grandchildren (who have forgotten all about the danger), and the cycle begins again.

Sefidabeh is an old settlement, a stop on one of the trade routes that crossed Iran, but it never grew large because of its unpromising location. But other villages that arose for similar reasons didn't remain villages.

For instance, three hundred kilometers to the southwest of Sefidabeh lies the ancient city of Bam. Another trading post on the routes between Persia and India, it was probably founded more than two thousand years ago during the Parthian Empire. Besides being a stopping point for caravans, Bam was known for growing cotton, dates, and fruit in fields irrigated by a series of channels that carried water from underground springs to the fields. Unlike Sefidabeh, Bam was able to expand from its original village settlement, and at the end of the twentieth century had a population of about ninety thousand.

But just like Sefidabeh, Bam owed its springs and its existence to a hidden fault. Toward the end of 2003 it is said that the inhabitants of Bam started to hear strange noises—distant booms they had never heard before. They concluded that some sort of secret weapons testing was going on in the nearby desert and thought it best not to inquire too closely. Had they realized the truth, tragedy might have been averted. In reality the fault, which ran past their city and had been locked shut by friction for centuries, was beginning to break apart.

In the early hours of December 26, the main earthquake struck with terrific force. As in Port-au-Prince the destructive force of the earthquake struck buildings that were so weak they offered little resistance. The main construction material was adobe, a sort of brick made out of mud, which was plentiful in the vicinity. Even the imposing fortress overlooking the city, the Arg-é Bam, a UNESCO World Heritage site, was made of adobe. I spoke later to an engineer who had visited Bam a few years before the earthquake. He described how in one street he had picked up a couple of loose bricks he saw lying on the ground and banged them together. Both turned to dust in his hands. Such weakly made houses could never have stood up to strong earthquake shaking, and across the city houses collapsed in on the sleeping inhabitants. About a third of the population was killed and another third injured.

But Bam is not the only case of a village on a fault line that has grown: consider Tehran. In the ninth century A.D. Tehran was a small village nestled at the foot of the Elburz Mountains in northern Iran. The mountainscape to the north changes abruptly to a fertile plain; the line where the mountains seem to have been sheared off is another fault—the North Tehran Fault—and it too is characterized by spring waters that make settlement attractive. During the period of the Safavid dynasty in the seventeenth century, Tehran became a royal residence, and it was made the capital in 1795.

It remains the capital of Iran today, swollen to a population of more than twelve million in the greater metropolitan area—four times the size of Port-au-Prince and just as exposed to earthquake hazard. Historical records show that major earthquakes occurred along either the North Tehran Fault or one of the neighboring related faults in 855, 958, 1177, and 1834.[2] The population of Tehran was relatively modest even as late as 1834. Now that it is twelve million, what will be the impact of the next Tehran earthquake?

THE MEGADEATH

Seismologists and disaster experts have been waiting anxiously for the first megadeath earthquake, that is, the first earthquake in history that will kill more than a million people. We have come close already. The highest death toll on record is for an earthquake in Shaanxi province in central China in 1556 that killed more than 800,000. Then there was the Tangshan earthquake of July 28, 1976, which occurred during China's Cultural Revolution, a time when the authorities were not prepared to acknowledge to the outside world exactly what had happened, much less release a death toll. A suspiciously accurate figure of 655,237 appeared in a report published in the *South China Morning Post* in Hong Kong the following January. The official death toll acknowledged today is about 240,000, and while some seismologists still suspect the true figure may have been twice that, even a quarter of a million represents a fatality rate of about 25 percent, which is high.

Tehran is one example of a city where the million death calamity could occur, but it is not the only place. Many places evidence a consistent pattern of earthquakes (in Tehran, four earthquakes during a period of twelve hundred years) but a completely changing pattern of human exposure, as villages become cities and cities become megacities. For thousands of years the planet supported only a small population of humans, which started to increase seriously only after 1800. Since then population growth has skyrocketed.

The reason for this is simple. Before the advances in medical knowledge and practice that began early in the nineteenth century, childbirth was dangerous, and the chances were low that a newborn baby would survive for long. For the population as a whole to remain stable, the high rate of infant mortality had to be countered by a high birthrate. If a family had a large number of children, there was a reasonable chance that two might survive to adulthood and perpetuate the line. It was generally expected that some of one's children, perhaps even the majority, would not live for more than a year or two.

This pattern has led to a common misunderstanding, even amongst people who should know better, that is worth correcting here. It involves the meaning of the term *life expectancy,* or the average number of years a newborn baby can be expected to live. In medieval times this was about thirty. This has led to such occasional statements, in books written by otherwise intelligent people, as, "At the age of thirty-five, he was already an old man, given that the average age was only thirty." Not so. Imagine ten infants, six of whom die around the time of their first birthday, with the remainder living to normal human old age and dying at seventy. Their average age at death is $1 + 1 + 1 + 1 + 1 + 1 + 70 + 70 + 70 + 70$ divided by 10, which is 28.6. But none of them actually died at the age of twenty-eight. Averages can be misleading, particularly when they are dealing with very disparate things, in this case babies and old people.

In the nineteenth century improved hygiene and medicine drastically lowered the infant death rate in Europe, but people were still used to having large families to compensate for a high infant mortality rate. The result was a huge boom in population—children were surviving to adulthood at a rate that far exceeded what was needed to keep the population stable. This produced a large population surplus that could not be accommodated. But a solution was available: emigration. Those who could not find work or land to farm in Europe simply took off for the Americas or Australia or Africa. Eventually the European birthrate adjusted itself downward until it reached a level at which the population was stable once more. In the twentieth century the story repeated itself in the developing world: improvements in medicine brought the infant death rate down, but the birthrate remained high. The result was the same as in Europe: a huge population surplus. The difference was that no vast open spaces remained to emigrate to.

So what happens to the population surplus in the developing world? The rural economies of Africa and India can support only so many, so once again some sort of migration has to occur. Some of it is external, not in the form of colonization of undeveloped open spaces but as economic migration to developed countries. But much of it also takes the form of internal migration from the countryside to cities, as if the cities were not expanding enough already.

Thus the global population is increasing at a rapid rate, and the increase is disproportionately in the urban population of the developing world. Given that these internal migrations are generally of people with few or no economic resources, cities end up with a vast sprawl of overcrowded housing of poor quality.

The statistics show it clearly. Since 1960 the overall world population has increased from about three billion to six billion, whereas the urban population has increased from less than half a billion worldwide to about five times that, and rising. Meanwhile the size of the rural population has flattened out and is even declining. Developed nations account for only about a sixth of the world's population.

These vast concentrations of people in the cities of the developing world can be vulnerable targets for natural disasters. One example is Turkey. Since 1960 there has been a persistent pattern of people packing their things, moving away from the villages in the Turkish countryside where their families lived for generations, and moving to the industrialized northwest of the country in search of jobs and better living conditions. In response, a great deal of housing was thrown up around Istanbul, and to the east and southeast, to accommodate the influx. And "thrown up" is all too accurate a description of the sort of building that took place.

Typically the new urban migrants moved into small apartment blocks of five stories, built in a hurry and built as cheaply as possible, using poor-quality construction materials. Reinforcing rods often were any old bits of rusty iron that had been lying around—never mind that the Istanbul area is a well-known earthquake zone. Evidently the builders were more interested in turning a quick profit than building earthquake-safe housing.

The inevitable struck in August 1999. An earthquake at three in the morning reduced many of these blocks to rubble, trapping and killing the occupants, who were mostly in bed and asleep. One striking photograph taken a few days later showed a local mosque with its slender minarets still standing while the housing all around it was in ruins. The official death toll of the Izmit earthquake was seventeen thousand; the Romanian seismologist Vassile Marza, who made a study of fatality counts from major earthquakes, believed the true figure was at least twice that.[3]

But not all such densely populated cities are in danger of suffering massive fatalities as a result of an earthquake. The same pattern of internal migration can be observed in Brazil. Especially since the 1970s, migration to its larger cities has resulted in large numbers of people condemned to living in chaotic shantytowns known as *favelas*. São Paulo is a city of remarkable contrasts. It is the world's seventh-largest city, with a population of more than ten million. While the richest inhabitants enjoy a high standard of living in the best neighborhoods, the city is surrounded by a ring of favelas for the indigent migrant population. But whatever problems the city might face, earthquakes are almost certainly not one of them. Brazil is conspicuously free of earthquakes.

Earthquake risk is a term that is often used loosely, but it has a precise meaning for seismologists. A lay person might talk about the risk of an earthquake occurring, but the technical meaning of seismic risk is the chance that actual loss—damage or death—will occur. This risk has three components: hazard, exposure, and vulnerability.

Hazard refers to the chance that an earthquake or, to be more specific, shaking will occur. The effects are quite different if an earthquake happens right under your house or a hundred kilometers away. A small earthquake at close range may be more dangerous than a larger earthquake far away. So it's never enough just to ask if an earthquake will occur. What matters is how badly the ground shakes. *Exposure* is a simple matter: Can anything be damaged? The early 1980s seemed for a while to be a period with fewer earthquakes than usual (which, except for seismologists, no one noticed). Certainly several years went by without any great earthquakes, until a massive earthquake occurred halfway between New Zealand and Antarctica in 1983. It was strongly felt on the remote McQuarrie Island, whose population is largely penguins. It went unnoticed by most of the world's media. If there is nothing to be damaged, no disaster occurs.

Last is *vulnerability*, which is a measure of how strong or weak buildings are. If the houses around Izmit had been better constructed, fewer would have collapsed and the death toll would have been much lower. As a rule an earthquake in California will cause less damage than an equivalent earthquake in Iran, because the buildings in California are more resistant.

It takes all three factors for a disaster. So far as seismic safety is concerned, it doesn't matter too much how badly Brazilian favelas are constructed. If the earthquake hazard is low, then the risk is low as well.

Similarly, although the Kamchatka Peninsula in the extreme east of Siberia is often subject to strong earthquakes, hardly anyone lives there. There's next to no exposure.

Looking at it in terms of hazard and risk, that "earthquake disaster potential" scale is a crude measure of risk; crude because it is simply one person's subjective rating. The author of it wanted to make the point that he could give some places high scores where vulnerability is high, but more specifically, he wanted to point out places where vulnerability is high specifically because the hazard has been underestimated.

If one looks at the global distribution of deaths from earthquakes in the last four hundred years, a remarkable picture emerges: fully 85 percent are concentrated in a roughly east-west zone running from Portugal to Japan. This area accounts for a mere 12 percent of Earth's surface but, with a combination of strong earthquakes, high population, and poor housing, it forms the world's greatest confluence of hazard, exposure, and vulnerability.

Earthquake hazard is constant. But just because earthquakes occur at the same rate, it doesn't mean that the risk is unchanging. If cities continue to grow, the potential for disaster increases, because the exposure is growing. As the world's urban population rises, the danger is that earthquake disasters will become more frequent and more severe.

This book is about how to stop that.

The chapters fall naturally into two parts. To understand the problems presented to society by earthquakes, it is necessary to know something about earthquakes themselves: what they are, what they do, and why they happen. We need a vocabulary to talk about them and to know something about how they are recorded, measured, and studied. So a little elementary seismology is worthwhile knowledge. Thus Part 1 is a guide to earthquakes that reaches from the seismologist's laboratory to deep inside the planet on which we live.

Part 2 considers what can be done about earthquakes. Can they be predicted? Can they be controlled? What can be done to make earthquakes less destructive—and who should be doing it?

Is there anything *you* can do?

2

WHAT *IS* AN EARTHQUAKE, ANYWAY?

A DEALER IN OLIVE OIL IS ABOUT TO ENJOY HIS DINNER early one evening sometime in the first years of the sixth century B.C.

Business has been good. He's more or less cornered the market in this part of Asia Minor, and as a result he has a pleasant country estate in the hills above the estuary of the river Meander; his property looks out over the blue Aegean in the distance. And this evening, as the sun slowly sinks into the sea, he is sitting on the vine-shaded terrace, a cup of wine in his hand, as he listens to the sounds of his slaves preparing the evening meal in the kitchen. It's quite a pleasant evening; the heat of the day has faded to a comfortable warmth, and the few small flecks of cloud in the sky show pink in the rays of the setting sun.

He takes a sip of wine (which, being a civilized Greek, he drinks watered) and reflects on how good life is.

Then he becomes aware of a low rumbling sound in the distance. It sounds like thunder, but the sky is blue. Then he notices another sound—the cups on the table before him are clinking against each other. In fact, everything on the table is trembling slightly, and the bunches of ripening grapes hanging from the vines on the trellises are moving to and fro.

Just as he wonders what in Hades is going on, his chair starts rocking violently. The table is rocking, too; the jugs tip over, spilling wine all over the place; plates and knives go skittering off the table and onto the ground. The rumbling of thunder is now a roar. He tries to get up out of the chair, but this is not a good idea; the chair tips over, and he lands on his back with a painful thump. The frightened slaves in the kitchen shriek as they watch amphorae overturn and break, and kitchen utensils rain from the shelves onto the floor, many of them breaking. An unholy clattering and crashing is punctuated by screams of alarm. As the olive oil dealer tries to stand, the solid flagstones of the terrace feel like the deck of a ship in a storm, pitching and bucking.

Suddenly it is all over. The ground beneath him is stable once more and the air is still, except for flocks of birds that took flight in alarm. He scrambles to his feet. The terrace is a mess. At least no one has been badly hurt, although his slaves are still trembling with shock and fright; in fact, he himself is still trembling from the unnerving experience. Most of the damage was done to crockery, though he can see a large crack in one of the house walls that wasn't there before. Fortunately it seems to be only a plaster crack and probably can be repaired fairly easily.

What on earth just happened? What was it he witnessed? At least he has a word for the experience: *sismos,* or, as we know it today, *earthquake.* But what, exactly, is this sismos?

At the simplest level the word refers to the phenomena he just witnessed—noise and shaking, coming as if from nowhere. But it's just a description; it doesn't tell what it is, or where it comes from. What sort of mechanism can suddenly make the ground shake through invisible force?

The obvious answer is the gods did it. This is what the slaves in the kitchen are thinking, and in fact they know which god—Poseidon. Poseidon is god of the sea; the sea shakes in a storm; so when the land shakes, that's obviously Poseidon as well. He's upset about something, and maybe it would be a good idea to visit his nearest temple (there's one down the hill in Miletus) as soon as possible and make an offering. Gods do everything through magic, so there's not a lot of point in inquiring further.

However, our olive oil dealer is an educated man and skeptical about this. Certainly it was divine power that created all things in the first place, but one would never get anywhere by superstitiously attributing everything that happens in the world to magic forces unleashed by gods who are curiously averse to showing themselves in public. For instance, take solar eclipses. The superstitious believe that a monster is eating the sun—but the merchant knows that they occur when the moon moves in front of the sun, and he can prove it, because from that knowledge he has been able to predict an eclipse.

So an earthquake is probably something similar, an event with a natural process behind it, rather than an unknowable divine magic unleashed when Poseidon waves his trident about. The problem is, first, what is that natural process, and, second, how can one find out what it is? Starting from nothing, it's not simple to see how to conduct such an investigation.

Astronomy is relatively easy. You can make observations of the movements of the sun and moon and planets every day and every night, thereby acquiring plenty of data in which to find significant patterns— and the Egyptians were doing just that long before the Greeks took an interest. But what sort of data will unravel the earthquake problem?

The olive oil dealer sets his chair upright, sits down, and tries to think about the experience that he just went through. He can remember, more or less, the sequence of events, even though he was so unprepared and surprised that it was hard to observe carefully and rationally. But that was the first time he ever experienced anything like that, so it's quite likely he may not have a second chance to see it and take better notes.

He has heard stories of other earthquakes that have happened around the Aegean—would it help if he could make a list? Would it be useful to know that during the thirty-fourth Olympiad an earthquake occurred in Sparta, and, if so, how would that be useful?

And what might be relevant? Was the position of the planets special in any way? What about the weather? Could it be related to that?

So far as anyone can tell, there were just as many geniuses in the sixth century B.C., proportionately, as there are today. Their handicap was that they could not benefit from the acquired knowledge of the centuries that we have at our disposal today. A twenty-first-century scientist starts from the position of being able to take for granted a huge amount of basic information about the natural world, and can move on from there into specialized fields of inquiry.

But the ancients had to start from scratch. So with no previous knowledge, how do you work out what an earthquake is? Let's go back to the olive oil merchant and give him his proper name, which was Thales.

WATER OR WIND

We don't know if Thales ever had his dinner ruined in the way I described, but he was interested in earthquakes. In fact, he was interested in everything. And he developed a theory of everything, which went roughly like this:

The world is full of a lot of different substances. But it's unlikely that it was created like that. More likely the world was created out of one element that later diversified. So what could that primary substance be? The obvious choice is water. Heat water and it turns into steam, a sort of air. A watery swamp, when it dries out, turns into earth. So water can turn into air or earth. On the other hand, earth never turns into water or air. Air might turn into water when it rains, but it never turns into earth. Fire never turns into anything at all. So—water is the primary substance.

Furthermore, there is a lot of water. So far as ancient Greeks knew, there is more sea than land in the world. Also, the land is higher than the sea. So the land sits on top of the sea, not the other way round. So probably there is a big primordial ocean, some of which has turned into land that sits on top of it.

Now, if this is the case, a theory of earthquakes comes rather easily. When a storm occurs, the open sea becomes agitated and waves form. Probably something similar happens to the hidden sea as well. In my story Thales noticed that the floor of the terrace was bucking like the deck of a ship in a rough sea, and, as a well-traveled man, he was able to draw a connection.

So if the land moves like the waves of the sea, the obvious reason is that the primordial sea is creating waves, and these are moving the land on top of it, just as the waves of the Aegean toss a boat to and fro. Not only that, but Thales was well aware that new springs sometimes form after an earthquake—clearly a case of disturbed water being forced up through the earth. There we have it—a theory that is logical and fits nicely with observations.

And so we can credit Thales of Miletus with the first-ever scientific theory of earthquakes. And, curiously, in some respects it was actually better than any other theory for the next two thousand years.[1]

As with many other ancient Greek writers, not a single copy of any of Thales' books survives. What we know of his work comes from the writings of later authors, notably Seneca.

Seneca was a Roman author known principally today as a playwright. But he also tried his hand at compiling an encyclopedia of the natural world, which he called the *Naturales Quaestiones*. Book 6 of this work is entirely devoted to a discussion of earthquakes, inspired partly by reports of a strong earthquake that occurred in southern Italy on February 5, 48 A.D. This earthquake caused considerable damage in Pompeii and Herculaneum, towns that thirty-one years later would be overwhelmed by the eruption of Vesuvius.

As a skilled rhetorician, Seneca composed his account in a fine and florid style, starting by running through the various theories of earthquakes that had been advanced by previous writers and then largely skewering them. He makes short work of poor Thales. If the earth were floating on water, earthquakes would not be rare events; on the contrary, the earth should be pitching about almost continually. Furthermore, if the earth is rocked by underwater waves, the whole floating landmass should move to the same degree as it tilts back and forth. This manifestly doesn't happen. The earthquake in Campania was not felt in Rome, so it must have originated from a purely local cause.

Seneca finally comes down in favor of air as the cause of earthquakes, and since this idea is largely associated with Aristotle, we should next turn to him.

Aristotle, of course, was one of the greatest of Greek philosophers—or perhaps, one should say, Macedonian philosophers, as he was born (in 384 B.C.) in Stageira, to the east of what is now Thessaloniki. Though he spent twenty years of his working life in Athens, he subsequently returned to Macedon at the invitation of the king, Philip II.

Aristotle's writings cover a wide range of subjects; this was long before the idea of specialist learning. Generally speaking, a learned man was expected to be learned about everything—which was easier in a period when not so much was known about anything. It was Aristotle's good fortune that many of his writings survived, unlike those of most of his forebears; his manuscripts were abandoned in a cellar for a few hundred years before being rediscovered and brought back to Athens. This was important, because Aristotle's development of a system of logic and scientific reasoning would have a huge influence on the development of science itself in the centuries to come.

Like Seneca, Aristotle starts by surveying previous theories in order to refute them. He skewers the work of various early Greek philosophers, not including Thales but including another Miletan, Anaximenes, who had tried to explain earthquakes as caused by the weather. Heavy rain makes earth sodden and it loses cohesion; therefore it falls apart and produces earthquakes. Then, again, earth cracks and breaks up in times of drought. So earthquakes should occur in times of drought or heavy rain, reasoned Anaximenes.

Aristotle thoroughly tears this apart. If the earth is constantly breaking up, with bits breaking off, the land should be sinking as a result—but this doesn't happen. Also, earthquakes often occur in places where there is neither drought nor heavy rain. It seems that Aristotle had collected some records of where and when earthquakes had occurred and put them to good use. Last, he points out that what breaks stays broken, so as the earth gets more and more broken by successive periods of drought and heavy rain, earthquakes should become less frequent. That also doesn't happen, so the theory doesn't work.

This is scientific reasoning at work, not just the application of logic, since Aristotle evidently had some data on past earthquakes that he was using to inform his reasoning. This becomes evident when we see Aristotle's own ideas. He presents several strands of evidence:

1. Earthquakes occur mostly at night; if by day, then about noon.
2. Earthquakes tend to occur in places where there are strong sea currents, many caves, or hot springs.
3. Earthquakes usually occur in spring or autumn.

None of these are actually true, so it would be fascinating to know exactly how much information Aristotle had.

Aristotle's basic reasoning is that since earthquakes are violent, they must originate with air, the most mobile of the four classical elements. A strong wind moves much faster than water, fire, or earth can ever hope to. Aristotle's reasoning went like this: rain soaks into the earth, the sun warms the earth, the moisture evaporates, the evaporation causes a wind inside the earth, the rush of the wind causes earthquakes. "We must suppose the action of the wind in the earth to be analogous to the tremors and throbbings caused in us by the force of the wind contained in our bodies," he writes.[2]

Aristotle suggests that these belches of air are more likely to occur when the weather is calm, and night is calmer than day, and noon is the calmest time of day. Hence earthquakes tend to occur at night or at noon. Earthquakes must occur more in spring or autumn, because the summer is too dry (not enough moisture to evaporate) and the winter is too cold (not enough heat for evaporation). Evidently where the earth is full of caves, the underground winds have plenty of passages to blow about in, and where sea currents are strong, these compress the air and make it more violent.

Aristotle also draws attention to the rumbling sound that comes with an earthquake: What could that be but the roaring of winds as they are forced through narrow caverns? And he is ready with an explanation for the formation of springs noticed by Thales: subterranean winds must force underground water to the surface. After all, wind drives waves on the sea, not vice versa.

Aristotle also notes that a strong earthquake is usually followed by many weaker ones—what we call aftershocks. This stands to reason, he says. Given that the earthquake is caused by the escape of trapped winds, it's unlikely that all the wind would escape at once. After the main blast a bit more follows, and then some more, until it is all released.

ARISTOTLE SAYS . . .

Theological explanations of earthquakes did not stop with Thales' slaves making offerings to Poseidon. As Christianity spread across

Europe, a distinctive amalgam of science and religion developed, in which the works of Aristotle became elevated almost to the status of holy writ. It began with Albertus Magnus in the thirteenth century; his interest in science was extensive and often based on observation. He absorbed the work of Aristotle and reinterpreted it from the perspective of a Dominican monk. The writings of his pupil St. Thomas Aquinas became a cornerstone of Christian doctrine, and in this way Aristotle's writings on the natural world achieved a status of nonnegotiable orthodoxy.

However, if Aristotle could have seen what became of his work in medieval Europe, it is not certain he would have approved. Aristotle's aim as a scientist was to use empirical evidence to infer natural causes. The medieval view, on the other hand, was dualistic. First and foremost, earthquakes were sent by God. However, God does not work by magic (as Poseidon might have) but by natural causes. If earthquakes are caused by winds forced through the inner cavities of the earth, God is deciding when and where the winds get forced.

So from the medieval point of view, if you want to understand earthquakes, you should really consider their theological significance, which is either (in the case of a major earthquake) to smite bad people or (in the case of a lesser earthquake) to warn people that if they don't mend their ways, more trouble is in store.

Since the Church had already decreed that Aristotle was right, disputing that view was dangerous. Science became the practice of writing books that agreed with Aristotle. Any argument could be settled with the phrase "Well, Aristotle says . . . "

The only way this stranglehold could be broken was for Aristotle to be found completely wrong on something, opening up the rest of his work to dispute. That breakthrough was achieved by Galileo in his work on astronomy in the early seventeenth century. Once it was proved and accepted that the old Aristotelian cosmogony of the sun and stars revolving around Earth was wrong, then the whole Aristotelian edifice was up for grabs, and it became possible to question other parts of it without running the risk of being tried by the Inquisition.

AN UNLIKELY EARTHQUAKE

The scientific revolution of the seventeenth century was characterized by an enthusiasm for learning through experimentation. Many amateur scientists of the day found an outlet in scientific societies where like-minded scholars could meet and exchange ideas and findings. One such was the Royal Society in London in 1660; in its early days its membership included the likes of Robert Boyle (famous for his advances in chemistry); Robert Hooke the inventor; John Flamsteed, who became the first Astronomer Royal; and John Wallis, a mathematician and cryptographer who, among other things, was the person who introduced the ∞ symbol to represent infinity.

And then occurred an earthquake unlikely to find itself a place in the history of seismology, the earthquake of January 29, 1666. This was no great disaster; indeed it was a trivial and insignificant event, strong enough to rattle windows across two English counties but no more. Wallis, sitting in his wood-paneled Oxford study, candlelit in the gloom of a snowy winter evening, had an experience very different from Thales'—a faint rumbling sound that Wallis at first took for carts in the street below, and that was it.

Boyle himself had a similar experience. That evening he was riding out to dinner with friends at a country house a few miles outside Oxford. The weather was foul, wet and cold, as he urged his horse to a gallop along a dark country lane. Soon after he finally arrived at his journey's end, doubtless while warming himself by the fire, he noticed a curious trembling of the whole house. On finding that all the other occupants had felt it too, he realized at length what it was he must have experienced.

This earthquake might have been a trivial event by world standards, but it was an earthquake nonetheless, and for the savants of Oxford, who had never felt one before, it was grist to their mill. Everything was interesting—even small earthquakes. So both Boyle and Wallis set out to collect what information they could, asking friends in neighboring villages for reports.

This was truly science in action—they were carefully assembling data that they could put together and try to interpret—and the data were then published for all to share.[3]

As it happens, not much in the way of new insights about earthquakes came about from their studies—hardly surprising, given the limited information that could come from such a small earthquake. But what was significant was the manner of investigation. In particular, they were not just interested in the sort of descriptive accounts Aristotle had collected. They wanted hard numbers from scientific instruments. This was a first.

Boyle and Wallis had nothing to record earthquakes with—and it would not have been much use if they had, since it would have been a bit late to use them. But what they could do was gather readings from weather instruments on the temperature and barometric pressure at the time of the earthquake. This was probably the first time in history that anyone had attempted to collect instrumental data in studying an earthquake.

Today that sort of meteorological data seem useless for studying earthquakes—but Boyle and Wallis didn't know that. Because they did not know exactly what an earthquake was, they couldn't be sure whether it was something that happens deep in the earth or something that happens in the atmosphere. After all, if an earthquake sounds like thunder, maybe it is a form of thunder. Perhaps some electrical charge like lightning forms in the atmosphere and communicates itself to the ground in the form of shaking. If that were so, meteorological data would be relevant, so collecting them would be a good idea.

In fact, there were other grounds for thinking that earthquakes might be aerial in origin.

ABOVE OR BELOW?

Earthquakes were in the news in the early 1690s. In 1692 the earthquake that destroyed Port Royal in Jamaica on June 7 was followed in September by one of the strongest earthquakes in northern Europe—it originated in the Ardennes, near Verviers, and shook much of northern France, the Low Countries, and even England. Then, on January 11, 1693, a strong Italian earthquake struck eastern Sicily, wiping out the town of Noto and causing heavy destruction in Catania.

The last of these inspired the astronomer John Flamsteed to put pen to paper and sort out the earthquake question.[4] He made a number of observations. First, he said, earthquakes always happen in calm seasons (not true, but his data were limited). Second, a noise in the air is always heard just before the shaking starts. Third, earthquakes can be felt at sea. An earthquake at sea felt the same to the crew as the vessel running aground (although there is nothing visible disturbing the sea surface), and Flamsteed had a number of reports of this. Fourth, some earthquakes affect large regions (like Sicily and southern Italy), whereas others are felt only over a small area (like the one at Oxford in 1666). Finally, and tellingly, he noted that the reports of the 1692 earthquake in London mentioned that it was felt most strongly on upper floors and was hardly felt at all in the street or in cellars.

This gave Flamsteed the evidence he needed to show that earthquakes could not be caused by underground shocks or explosions. If they occurred underground, earthquakes should happen in stormy or rainy conditions just as much as in calm conditions. Also, a noise in the air means something is happening in the air, not in the ground. Further, how could an underground shock cause a ship floating on the sea to shake, when the sea itself remained calm? And if an earthquake can be felt across a large region, if it were caused by explosions in subterranean caverns, it follows that the cavern system must extend everywhere, which seems inherently improbable. Last and conclusively, a shock originating underground would be felt more strongly at ground level and least strongly on the upper floors of buildings, the reverse of what is observed.

Therefore earthquakes must be atmospheric explosions.

This is a wonderful example of scientific reasoning, working from observation to conclusions. Unfortunately it's all completely wrong. But it demonstrates beautifully the process of trying to arrive at a theory of earthquakes from all the information that could be collected.

In the history of science, fame has gone to those who got it right, but being instructively wrong is a good second best. Flamsteed actually did get one thing right—the idea of extensive caverns under the whole of Europe is implausible. As for the idea that earthquakes happen only in calm weather—well, if that were true, his inference was reasonable. But he had too small a sample of earthquakes at his disposable; they do in fact occur in all sorts of weather conditions. For the rest? Well, he was lacking one crucial piece of the puzzle, one that would not turn up until midway through the next century. And it would take one of the greatest earthquake disasters to provide it.

THE LESSON OF LISBON

Place: Lisbon, Portugal. Date: November 1,
1755, the morning of All Saints' Day.

The first indication that anything was amiss on this, one of the most solemn dates in the Catholic calendar, was a rumbling sound, like heavy traffic. The ground shook, there was a pause, and then the ground heaved with terrific violence for more than two minutes, bringing buildings crashing down all over the Portuguese capital. Then a third shock finished off many buildings that had been left teetering by the second, sending up huge clouds of thick, choking dust.

Then came the fires. On a November morning the houses would have been heated by lit stoves or open fires. After the buildings collapsed, the ruins easily caught fire. The fire spread, taking hold of the royal palace, the opera house, and many other buildings that had survived the shaking. A cold northeast wind fanned the blaze, which would continue burning unchecked for almost a week.

But more was to come. Many survivors, looking to escape the raging fires, sought refuge on the quayside, down by the estuary of the river Tagus. That the sea seemed to have retreated was clearly strange, but with the city in ruins and fires out of control, people thought they had enough other things to worry about. About eleven o'clock they discovered how wrong they were. There was another roaring sound, and then huge waves suddenly swept up the estuary, throwing ships around like toys, crashing over the quayside, and surging up into the ruined city. In the low-lying parts of Lisbon, any buildings that survived the earthquake and fire were finished off by the tsunami.

Europe reeled. Though the apogee of the Portuguese Empire had passed some centuries earlier, Lisbon was still one of the richest capital cities of the world, and its destruction came as a terrible shock. More shocking still were the date and time. The pious inhabitants of Lisbon who packed the churches to hear mass on this holy day were all killed as the churches collapsed about their heads. On the other hand, the sinners who skipped mass and took their pleasure in parks or gardens survived.

That earthquakes were sent by God to punish the wicked was still an established article of faith. How on earth could that be reconciled with a disaster that seemed to target God's faithful, that destroyed his houses of worship and left the reprobates alone? It made no sense at all. The earthquake struck not just at the buildings of Lisbon but at the core belief about how God operates in the world. Trust that everything that occurred in the world was divinely ordained crumbled as Lisbon's churches collapsed.[5]

Another explanation had to be found.

After the physical shock wave had traveled across Europe, disturbing lake waters as far away as Scotland and Scandinavia, a sort of intellectual shock wave followed it. Old certainties were dashed, people wanted to know what they should believe, and Europe's intelligentsia set about trying to provide some answers. A rush of books and pamphlets about earthquakes followed.

One philosopher who took up the challenge was Immanuel Kant. Kant was one of the most famous philosophers of the eighteenth century, and his writings have cast a long shadow over philosophy ever since. He was born in the city of Königsberg, in East Prussia (now Kaliningrad, in a small enclave of Russian territory), and spent his entire life there. In 1755 he was just thirty-one; his major works, such as the "Critique of Pure Reason," lay well in the future.

Kant produced three essays on earthquakes in which he strongly rejected any supernatural agency. "We stand with our feet on the cause," he remarked.[6] One detail that drew his attention was the effect on lake waters far removed from Portugal on the day of the Lisbon earthquake. How could an earthquake reach across such large distances? His answer was to re-invoke Aristotle's subterranean caverns, but this time with a twist. These underground caverns must run parallel to mountain ranges and large rivers. These would serve to guide earthquake shocks, and in a nice nod to observational evidence, he pointed out that swinging objects in Italy moved from north to south, parallel to the Apennines, while in Portugal they swung east-west.

Not satisfied with pent-up winds, Kant proposed chemical explosions as the ultimate cause. Earlier writers had reported that in the presence of water, a mixture of equal parts sulfuric acid and iron would spontaneously combust. He envisaged a system whereby hot compressed air from an underground explosion would course through the cave system, eventually finding egress through volcanoes. The underground combustion would then be stifled for lack of oxygen; air would rush in from outside to fill the empty space, causing another explosion, and so on, in a sort of gigantic respiration cycle. The caverns, Kant thought, were like the lungs of the planet.

If the explosion were large enough, it would cause an actual tilting of the earth across large distances. This would not produce a noticeable effect but would be enough to cause lake waters to slop about, as was observed at Loch Ness, among other places, during the Lisbon earthquake.

Furthermore the theory had practical consequences. Evidently, Kant thought, Lisbon suffered because it was built alongside the river Tagus, maximizing the city's exposure to the passage of the earthquake. Building cities parallel to mountains or major rivers meant that they were running alongside the caverns where a future earthquake would occur, increasing the damage they would suffer, so such sites should be avoided.

Last, he was pleased to note that since East Prussia lacks any mountains, he could consider that, tucked away in Königsberg, he was pretty safe from earthquakes. At least there he was right—but not because of the flatness of the countryside.

Kant is a good example of someone who was trying to bring the subject of earthquakes into line with other contemporary advances in science. He knew about advances in chemistry made by his contemporaries, and he was able to draw on that knowledge to come up with hypotheses that were at least credible. But, like Flamsteed, he was missing a critical piece of the puzzle. That piece was about to be supplied.

ACTION AT A DISTANCE

History has not been kind to John Michell. Although he was one of the most original scientific thinkers of the eighteenth century, he is hardly remembered today. He was born in 1724 and entered Queen's College, Cambridge, at seventeen. After being elected a Fellow of the college after graduation and appointed Woodwardian Professor of Geology, he gave up both fellowship and professorship on the occasion of his marriage in 1764. He left Cambridge for the quiet life of a country rector, first in Hampshire, and then at the little Yorkshire village of Thornhill, where he died in 1793. It seems he lived a life of little incident but quiet contentment.

His intellectual life, however, was varied and remarkable. He invented a way to make artificial magnets, showed that magnetic force decreases as the square of distance, and paved the way for the first accurate measurement of the mass of the earth. In astronomy, among other advances, he made observations on stellar parallax that allowed him to make an accurate estimation of the distance from the solar system to the nearest star. After studying the effect of gravity on light, he postulated that a heavy enough stellar object could prevent light from escaping—the first description of what we now know as a black hole.

Michell, like Kant, was inspired by the disaster at Lisbon to consider earthquakes, and he brought a greater lucidity to his reasoning than anyone theretofore. He outlined his ideas in an address he gave to the Royal Society in 1760.[7]

The bugbear that had plagued Flamsteed, Kant, and their contemporaries (and, of course, I have passed over many other names that could be mentioned) was that of action at a distance. How did the force of an earthquake transmit itself from one place to another? There had to be some sort of medium for it, and the only possible answer seemed to be air. Whether it was the movement of gas in the atmosphere (for Flamsteed) or underground (for Kant), it had to be the movement of air that carried the force of the shock. Wrong answer.

Michell provided the right answer: elastic waves. It might seem strange to think of the earth as being elastic—if you pick up a lump of rock, it is not exactly bendy or stretchy. But actually there is no such thing as a completely rigid body, and across large enough distances you will see elasticity even with the eye. Imagine a stone rod the thickness of a pencil and twenty feet long. Would you be surprised if it curved slightly if you held it out by one end? I don't think so.

Michell's insight was that a sharp force acting on the earth, say, a compression, would travel through the solid rock as a wave, and the compression would be followed by a dilation, and then another compression, and the net effect would be a vibration that one could feel at the surface. Rather like sound waves.

It had long been known (since Leonardo da Vinci, in fact) that sound travels as waves. But Michell proposed that earthquakes do too. And that suddenly illuminated another observation about earthquakes—the loud rumbling that precedes them. "If these alternate dilations and compressions should succeed one another at very short intervals, they would excite a like motion in the air, and thereby occasion a considerable noise." Or, to put it another way, some of the wave energy bleeds from the earth into the atmosphere, where we perceive it as sound.

Now we can return to Flamsteed and dispose of those fallacies. Earthquakes only in calm weather—not true. Noise perceived in the air—explained. In fact, Michell went even further and argued that since noise and vibration arrive at about the same time, the speed of earthquake waves must approximate the speed of sound. This overlooks that the leakage of waves into the air occurs everywhere, not just at the source, but it remains the first attempt to estimate seismic velocity.

Earthquakes observed at sea cause no difficulty, since shock waves can travel quite happily through water without disturbing it. Ironically Kant grasped this but failed to see the general significance of it. Extensive underground caverns, of course, are no longer needed.

Last, what about earthquakes being felt more on upper floors? Michell does not address this, but the answer is straightforward. A shock wave traveling through the earth hits a building at its base and sets it vibrating. The foundation of a house is fixed in the ground and can't move to and fro. The roof of a house is not anchored to anything external to the structure. Therefore the building acts like an inverted pendulum, where the top has much more freedom to sway about. Hence earthquake shaking usually seems much stronger the higher up in a building you are.

Michell then went on to invent single-handedly the whole concept of earthquake location, at least in outline. Suppose you have two observers at different locations. One feels an earthquake shock coming from the southwest. The other feels the same shock, but it is coming from the southeast. Draw a line southwestward from the first observer and southeastward from the second, and where the two lines cross is the origin of the earthquake.

Alternatively, if the shock arrived at the first observer before it arrived at the second, the origin must be closer to the first observer. With enough observations one would be able to pinpoint the origin with increasing accuracy. However, as Michell realized, human powers of sensing direction or timekeeping were hardly up to the task, and he acknowledged that it would be virtually impossible to get sufficiently accurate information in the event of an earthquake. It would take the invention of proper recording instruments, and we will meet these ideas again in Chapter 4.

Michell was way ahead of his time. The mathematics of the propagation of elastic waves in solids had to wait for the work of Siméon Poisson and Augustin-Louis Cauchy in the 1820s, and even they were well ahead of observational data. And as so often happens when someone anticipates the ideas of a later era, Michell was largely ignored by those around him, at least so far as his writings on earthquakes were concerned.

When it came to the cause of earthquakes, however, Michell went off beam. Like many before him, he noted that there seemed to be a connection between earthquakes and volcanoes. He supposed that there existed great subterranean firepits of burning material (perhaps coal or shale). The rock strata above one of these firepits would be arched upward by the heat. Periodically bits of it would collapse, water in the rock would turn into vapor instantaneously as it fell into the fire, and this would produce a concussion that would engender an earthquake.

Not true, but it did seem to explain why earthquakes seemed to happen repeatedly in the same place; after a roof fall would come a recuperation phase in which the strata arched upward again before the next fall could take place.

No one really knew what produced earthquakes for a good reason—no one had ever seen one happen. Certainly people had seen the effects of earthquakes, but no one had ever seen the collapse of strata into a pit of flames three miles deep (Michell's guess at the depth of the source of the Lisbon earthquake), nor could they. Certainly not Michell, tucked away in the easy comfort of a Cambridge college, nor Kant, a recluse among the pleasant timbered houses of Königsberg. They could not see the cause of an earthquake; all they could do was cast about for hypotheses that seemed to fit the facts.

An earthquake would have to happen at the surface, where human eyes could see it, in order to determine what actually caused the shaking. And this was not going to happen in Europe, where earthquake sources were either too deep or offshore, or both.

Witnessing an earthquake actually at the surface would have to wait for another hundred years after the Lisbon disaster, and it would occur in a place as remote from Europe as possible—New Zealand.

THE GROUND BREAKS

The first European to reach New Zealand was the Dutch explorer Abel Tasman in 1642, hence the name: Zeeland being part of the Netherlands. But it was left untouched by Europeans until British and French expeditions started to use it as a stopping point in the late eighteenth century. The first European settlers were sealers, whalers, and loggers keen to exploit the rich forest cover. The sealers and whalers made camp and sailed on, but the loggers started to build permanent settlements.

Gradually, during the early nineteenth century, more settlers began to drift in—ex-convicts from Australia, missionaries from Britain, and, increasingly during and after the 1830s, settlers driven out of Australia by drought and economic depression. By 1840 the number of Europeans in New Zealand had reached a couple of thousand. These isolated pioneers, cut off by vast distance from their homeland, faced a harsh existence as they attempted to hack out a living from a primitive wilderness.

And they discovered something else—their new home was rather earthquake prone. The first indication came in 1843, when strong shaking, lasting as long as a couple of minutes, was felt all over the southern part of North Island, damaging houses and causing many landslides, one of which overwhelmed two colonists—the first Europeans to die in a New Zealand earthquake (of the Maori past, of course, we have no record).

In 1848 a similar shock struck the Marlborough region of South Island, causing damage on both sides of the Cook Strait. In Wellington many brick and stone houses were so badly damaged they needed to be rebuilt.

Then, in 1855, on January 23, an even more powerful earthquake occurred, close to Wellington at the southern tip of North Island. Despite the heavy shaking, the town escaped largely unscathed, as the wooden houses built to replace the brick ones damaged seven years previously proved to be much better at withstanding the earthquake motion.

What was astounding was the effect on the landscape. The whole countryside seemed to have been wrenched apart. Stream beds that had previously been straight now had a huge eighteen-meter kink in them, and in places the land appeared to have been lifted up violently on a large scale. This included part of Wellington harbor, where some jetties and quays were now so far above sea level as to be unusable. Elsewhere, cliffs up to three meters high suddenly appeared along a line that cut straight across the terrain.

Back in Britain, one man was particularly interested in events in New Zealand: Sir Charles Lyell.

In 1855 Lyell was recognized as the foremost geologist of his day, largely thanks to his enormously popular and influential three-volume work, *Principles of Geology*, which was already in its ninth edition. Born near Dundee in Scotland in 1797, Lyell trained at first to be a lawyer, but he found that poring over endless dusty tomes of case law proved seriously deleterious to his eyesight. But he was from a wealthy family and could more or less please himself as to what he did, so he switched to geology as a more congenial occupation.

In his writing and lecturing Lyell was a strong proponent of the principle of uniformitarianism, established by another Scottish geologist, James Hutton, in the late eighteenth century. Before Hutton it was widely believed that geological formations showed the effects of cataclysmic upheavals of the past—principally the great flood described in the story of Noah. Hutton showed that the evidence found by actually looking at rocks pointed in a quite different direction—that the landscape was shaped by slow, gradual processes that can be seen going on today—erosion wears rocks away, rivers transport the silt and sand to the sea, and the layers of sediment on the seabed are gradually compressed until they become compacted into rock.

Not surprisingly, this was a controversial idea, since these processes needed millions of years to create the current landscape, and this required the earth to be ancient—not six thousand years old, as had been calculated from the Old Testament. The acceptance of uniformitarianism in geology, against the word of the Bible, was in no small part the result of Lyell's writings.

One geological phenomenon that had to be explained was faulting. A fault is a dislocation across a set of rock strata—a plane of fracture where one can easily see that the rocks on one side have moved with respect to those on the other. The existence of faults was well known: they are frequently observed by miners, who naturally find it rather annoying to be happily excavating a rich seam of ore, only to find it terminating abruptly in a blank wall of some other rock. Miners eventually learned that if they dug up a bit or down a bit, they could often find the seam again at a different level and resume their work on it.

Why did faults occur? For Lyell the explanation had to come from some process that continues to operate today; it was no good to suggest glibly that everything got shook up and broken when God sent the waters to deluge the earth. Lyell argued that once a rock bed had been broken along a fault line, the two sides would no longer be connected and would be able to move independently. Thus further movement along the same plane would be relatively easy. In this way a series of modest slippages along the same plane could produce a huge displacement over time. What was lacking was observational evidence that this actually happened.

And now, thanks to informants in New Zealand who provided him with firsthand descriptions of the landscape changes on North Island in a series of letters, he had what he needed—direct observation of a fault moving.[8] And it moved in the course of an earthquake. There was no doubt about it. This was faulting in action—today—not in the distant past. "The geologist has rarely enjoyed so good an opportunity as that afforded him by this convulsion in New Zealand, of observing one of the steps by which those great displacements of the rocks called 'faults' may in the course of ages be brought about," he wrote exultantly.[9]

But one problem remained. Lyell had established decisively that fault movement and earthquakes occurred together. But which was the cause and which was the effect? It was not altogether clear. Did the movement of the fault produce the shaking, or did the earthquake motion wrench the fault apart? And how could one decide?

Association between earthquakes and faults had been suggested previously, but before Lyell, no one could ever prove it. To hypothesize that an earthquake was caused by the dislocation of strata would have been just as much a conjecture as Kant's spontaneous explosions.

So far we have traveled from Greece to England and from Italy to Portugal and thence around the world to New Zealand. For the conclusion it is necessary to move to the United States, to San Francisco, California, to be precise. The year is now 1906.

To some extent the time line of settlement in California echoed that of New Zealand: little European presence before 1800 and a boom in the 1840s, fueled in California's case by the gold rush. But by 1906 California was no pioneer outpost. San Francisco was a majestic and beautiful city of opulent houses for the wealthy, stately public buildings, and a fine opera house. And on the evening on April 17, 1906, no less a person than Enrico Caruso, the greatest tenor of his generation, was present to take the role of Don José in Bizet's *Carmen*.

After the performance Caruso retired to sleep in his suite at the Palace Hotel. Built in 1875, the Palace claimed to be the largest hotel in the western United States, if not in the world. At 5:13 the next morning, Caruso, along with practically the entire population of the city, was abruptly jolted awake. He got unsteadily to his feet as the floor of his room continued to rock like the deck of a ship and went over to the window. What he saw in the early morning light horrified him. Everywhere he could see, buildings were toppling, huge chunks of masonry detaching themselves and plunging into the streets in clouds of dust, as cries and screams echoed from all directions over the roar of falling stone.

Aided by his valet, the singer threw on the first clothes that came to hand and beat a hasty exit from the hotel. Already ceilings were collapsing, sending showers of plaster over everything. He made his way to Union Square, where other members of the opera company were congregating, together with many others seeking safety. For after the shaking another peril had appeared—fire. As in Lisbon in 1755, what was not thrown down would perish in flames.

The military quickly moved to take control of the stricken city. To the soldiers' dismay, they found that the earthquake had put the city's entire water supply out of commission. The fires were steadily advancing across the city, and hoses were useless.

The only defense seemed to be to contain the fires by creating firebreaks. Parties were assembled to grab what explosives could be found and then systematically blow up belts of houses across the city to create barriers to the further spread of the flames. Three things went wrong with this plan. What they needed was dynamite, and for the most part they had only gunpowder. The problem with gunpowder explosions was that they often set fire to the buildings being blown up, causing new fires. Then, to try to contain the fires as rapidly as possible, the firebreak teams set their explosions too close to blocks that were already burning. A blast would reduce a wooden house to a pile of matchwood, which would then catch light from the flames, spreading the fire rather than containing it. Finally, the teams were not careful about which buildings they chose to demolish. One turned out to be a warehouse containing a vast stock of spirit alcohol. The resulting fireball caused an entirely new major fire.[10]

California at that time did not have a reputation for earthquakes. It should have—European settlers first probing the area had felt several. But throughout the nineteenth century there was a quiet policy of hushing up any earthquakes that occurred—bad for business! If eastern tenderfeet got the idea that California was earthquake country, they might have second thoughts about moving there.

The 1906 earthquake was rather hard to hush up, but that didn't stop people from trying. OK, so there was an earthquake—but it was the fire that caused all the damage—and fires can happen anywhere, can't they? For a time it was official policy to refer to the events of April 18 as the "great San Francisco fire" and try to airbrush the actual earthquake out of history. It didn't work. The 1906 earthquake is perhaps the most famous in history. It is an icon, the archetype of earthquakes in the public imagination.

The scientific community in California mobilized at once. Professor Andrew Lawson, chair of the Geology Department at the University of California, Berkeley, approached the state's governor, George Pardee, and persuaded him to appoint a state earthquake investigation commission. The aim was to pool the contributions of all interested parties, including scientists from Stanford University, the University of California, Johns Hopkins University, and, of course, the US Geological Survey (USGS). Within three days the commission was established under Lawson's leadership.

One scientist on the commission was Harry Fielding Reid, a geologist at Johns Hopkins University who had traveled widely in Europe and North America. In 1902 he started collaboration with the USGS, operating one of the few earthquake-recording stations then existing in the US. His entrée with the USGS gave him access to detailed surveying data that had been collected before the earthquake. Herein he discovered an important difference between California in 1906 and New Zealand in 1855.

As with the 1855 earthquake, fault movement in California was exposed at the surface—spectacularly so, with fences and roads ripped asunder by many meters, although this time the movement was just horizontal instead of horizontal and vertical. But the displacements in New Zealand occurred across more or less wild countryside. The Bay Area of California had been extensively surveyed to a high degree of accuracy. Here was the opportunity to resurvey and find with precision how much displacement had occurred.

The results were fascinating. The closer a point was to the fault, the farther it turned out to have moved. And a curious pattern emerged. Reid found that if he took a straight line of survey points that crossed the fault line perpendicularly and considered the line between two

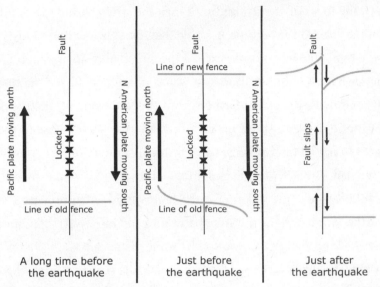

This diagram shows what happens before and during an earthquake with respect to lines of fence posts crossing a fault. Before an earthquake, posts far away from the fault move; during the earthquake, those closest to the fault move. It was this type of observation that enabled Reid to put forward his ideas on elastic rebound after the 1906 San Francisco earthquake.

points sufficiently far from the fault as not to have moved in the earthquake, what had been a straight line was now split into two curves: one curving up to meet the fault line, and the other, on the other side of the fault, curving downward. Yet before the earthquake these two lines had been straight and continuous.

Reid saw what it meant. Based on the measurements it was clear that the Pacific side had moved northward relative to the continental side. So whatever forces were involved, the western block was always being pushed north and the eastern block south. But something stopped any actual movement until the earthquake started—friction, obviously. Remember, though, that rocks have elasticity—they can deform and become compressed or expanded. The forces acting on either side of the locked fault would cause strain that would gradually build up until it was stronger than the force of friction holding back movement. When that happened, the fault would move with a jerk, releasing the strain in the form of energy. But until then the strain would gradually deform the rocks on either side, bending them in opposite directions. If one were to plant a line of stakes across the fault before any strain built up and sit back for a few decades to watch it, one would see the line slowly bend, with the stakes farthest from the fault moving the most, north on one side and south on the other, and the stakes close to the fault held back and not moving at all, producing a curve like a flattened and stretched-out S. When the fault broke, the two sides would snap in opposite directions: the stakes near the fault would catch up with the ones farther away, and the resulting pattern would be two straight lines on either side of the fault, now offset by some distance.

In the California case the USGS measured its survey points when the ground was already deformed by the build-up of strain, so the effect was the opposite. Instead of a curved line that becomes two straight ones as the fault abruptly moved, a straight line becomes two curved ones. Either way the overall pattern is the same—between earthquakes more movement occurs away from the fault than on it. Then, when the fault snaps, the ground on either side of the fault plays catch-up.

Reid called this elastic rebound. Imagine you are holding the end of a wooden ruler horizontally in your left hand, with your right thumb on the other end. You are moving your left hand slowly upward but keeping your right thumb motionless. The ruler gradually bends. Now remove your right thumb. The ruler springs back into shape with a *ba-doing!* and a good bit of vibration. It rebounds. This is not dissimilar to what happens along a fault. And now it was quite clear that rebounding faults release a lot of energy as compressional waves, much in the way that Michell supposed, but with a different originating cause. Faults produce earthquakes, not vice versa.[11]

And this, at last, was the complete answer to what an earthquake is: a violent movement of rocks that releases energy in the form of waves that spread outward at high velocity. The energy accumulates as strain when the strata on either side of a fault are deformed in response to constant slow-acting forces, until the force of friction is no longer able to stop movement from taking place. When the strain is too great for friction to bear, the fault breaks, the earthquake occurs, and then the whole process begins again.

But where do these forces come from? That is a different problem. After Reid the question "What is an earthquake?" had a definitive answer. Why they occur was still a mystery.

3

JOURNEY TO THE CENTER
OF THE EARTH

EARTHQUAKE. IN GERMAN *ERDBEBEN*. IN FRENCH *TREM-blement de terre*. In Italian *terremoto*. In all cases a compound: the earth shakes. By the end of the eighteenth century, the shaking part was reasonably understood, thanks to John Michell. But the earth, less so.

As we wander around the surface of our planet, we don't have a good view of what lies beneath our feet. There is soil, and if you dig down under that, you find rock. Dig farther down, in a mine, more rock. Or before mines were even invented, early humans explored deep underground using natural cave systems. What else was down there?

One persistent idea has been that somewhere underground was the abode of the dead—the underworld, Hades or Hell. The Greek philosopher Pythagoras, wrongly remembered today for the theorem that bears his name (its true discoverer is not known), even worked this into his own hypothesis on the origin of earthquakes—they were caused by mass meetings of the dead, deep down in Hades.[1] All those shuffling feet—

Except for those like Flamsteed, who thought earthquakes were an aerial phenomenon, an explanation of why earthquakes happen must lie in some aspect of the structure of the earth. Aristotle's ideas relied on extensive underground caverns. Kant's and Michell's required vast deposits of flammable materials.

The latter didn't seem so unreasonable, since coal mining was a well-established and important process for producing fuel even before the Industrial Revolution. And then there were volcanoes to explain. Clearly something underground was capable of generating fire at the surface, and it was widely believed that fiery volcanic eruptions were the result of the combustion of material—perhaps coal—underground. But it was still speculation. Caves could get you down only so far, mines likewise. It was one thing to ponder what might be deep down inside the Earth, but observing it, or proving it, was quite another matter.

What was known, and from a surprisingly early period, was that the world was round. Moreover, and notwithstanding popular songs, people didn't laugh at Christopher Columbus because he said the world was round, they laughed at him because he said the world was small. Columbus thought the world was so small it would be practical to sail west from Europe and reach China. He was wrong.

The size of the planet was actually calculated quite accurately in ancient times. Eratosthenes, a philosopher and mathematician who was also director of the famous Great Library of Alexandria at the end of the third century B.C., set himself the task of calculating the circumference of the earth. By observing the difference in the angles of the sun's rays at two points on the surface of the earth, Alexandria and Syene (now Aswan), he computed a value of 250,000 stadia for the earth's circumference. Allowing for uncertainty in the value of a stadium, the Greek unit of distance, this probably converts to about 46,000 kilometers, compared to a true distance of just more than 40,000 kilometers—an error of only 15 percent. For someone working within the constraints of third-century B.C. Egypt, this is a remarkably accurate result.

So with a radius of more than six thousand kilometers, a lot of planet is under our feet. Once the ideas of underground winds and vast, burning coal deposits had been discarded, it was necessary to find something else to explain why the earth could move so violently. It had to be something to do with how the earth is structured, but short of discovering a series of caves leading to the center of the earth, as in Jules Verne's science fiction, finding a way of observing the interior of the planet seemed an insurmountable obstacle.

TELEGRAPH HILL

If asked to name the most important date in the history of seismology, I think that most seismologists would cite April 17, 1889. Place: Potsdam, Germany.

Potsdam, a picturesque town just southwest of Berlin, flourished as the residence of the kings of Prussia. It is famous for the Sanssouci Palace and parks built by Frederick the Great in the eighteenth century. In the 1830s a telegraph line was built between Berlin and Koblenz, a distance of around five hundred kilometers. This was before the introduction of the electrical telegraph, and the line worked by means of a series of mechanical semaphore stations, one of which was built on a hill on the southern edge of Potsdam. It became known as the Telegrafenberg—Telegraph Hill—and by the 1870s it had become the home of the Royal Prussian Geodetic Institute.

Telegrafenberg is an eerie place, especially in winter. The hill is still thickly wooded, and as one wanders along the paths, strange brick buildings capped with massive domes loom up between the tall bare trees, a surreal and spectral landscape. In 1889 it was home to Ernst von Rebeur-Paschwitz, a scientist who was researching gravity by using a tiltmeter, an instrument designed to measure fine changes in gravity that result from subtle deformations in Earth that might be caused by tidal forces.

Rebeur-Paschwitz looks out at us from period photographs with a gentle face and intense eyes beneath a central part in his hair. A spindly moustache sticks out on either side of his face over a neatly trimmed beard. A contemporary once described him as "our incarnation of the ideal man of science. He had Darwin's lovable nature, as well as his modesty and utter carelessness of his own personal fame."[2]

On April 17 Rebeur-Paschwitz saw his sensitive instrument record something quite different from the usual slow wobble—the pen swung violently from side to side, leaving a sharp up and down spike on the chart where the pen had suddenly lurched. The initial swing was followed by a series of smaller rapid oscillations until gradually the instrument returned to its normal pattern. He had never seen anything like it. No one had. He did not realize it at the time, but he had just witnessed the first-ever recording of an earthquake that had occurred on another continent.

He was not entirely alone. Far away in Japan, in the offices of Tokyo Imperial University, Cargill Knott, a Scottish physicist who had taken up the post of professor of Physics and Engineering a few years previously, was chatting with a Japanese colleague, the geologist Sekiya Seikei. When the room started to tremble, the two men rushed to where several experimental recording instruments had been installed and were delighted to see the needles swinging wildly. While Knott and Seikei knew exactly what they were watching, it was a few months before the puzzled Rebeur-Paschwitz read a news report of a strong Japanese earthquake and realized that the time and day were consistent with his mystery recording.

The age of instrumental seismology had begun.

Rebeur-Paschwitz was well aware of the implications of his accidental breakthrough. Hitherto, if someone had wanted to study Japanese earthquakes, it meant going to Japan, like Knott. Now, with a suitable network of sufficiently sensitive instruments, one could record and study earthquakes from all over the world without leaving home. For the first time it would be possible to study earthquakes globally, not by collecting newspaper accounts that were often months out of date by the time they reached Europe, but virtually in real time. It would be possible to monitor the seismic pulse of Earth itself.

But time was already running out for Rebeur-Paschwitz. The first symptoms of tuberculosis were already evident, and a stay in Tenerife in search of a better climate did not stay the course of the disease. Undaunted, he threw himself into his work, laying out a plan for a worldwide network of seismic recording stations, with a central agency to collect and publish the data. He never saw it happen. He died in his parents' home in Merseburg, Saxony, in 1895. He was only thirty-four.

The global seismological network that Rebeur-Paschwitz foresaw would be realized by an English mining engineer, John Milne, who would set in motion a system that essentially survives to this day.

Milne was born in Rochdale, near Manchester, in 1850. After completing his training at the Royal School of Mines, Milne plunged into a life of travel in Europe and North America, and he joined an expedition to investigate Mount Sinai in Egypt. In 1875 he was appointed professor of Geology and Mining at the Imperial College of Engineering in Tokyo. Typically for Milne, he took the hard way of getting there, traveling overland through Siberia and making geological observations along the way.[3]

For the first few years in Japan, Milne busied himself with his academic duties and writing up the material he had collected during the previous decade. Earthquakes were not uppermost in his mind. But all that changed on February 22, 1880, when a damaging earthquake struck Yokohama and rattled the whole of Tokyo. It was not his first exposure to this force of nature; a weak earthquake had rumbled through Tokyo on the very first night Milne was there. But this time Milne's attention was caught.

At Milne's prompting a scientific meeting was called to discuss what should be done. By the end of the meeting the world's first seismological society had been founded, with Milne as Vice president.

Milne was not the only foreign scientist at the meeting; at this period of Japanese intellectual expansion, scientists from overseas were being sought to inject new ideas into Japanese academia. Along with the local Japanese, Milne now started working with two Scots, Thomas Gray and Alfred Ewing, on constructing instruments to record earthquakes. When Ewing returned to Britain, he was replaced by Cargill Knott, who would witness the recordings in Japan of Rebeur-Paschwitz's earthquake in 1889. The result of the labors of this little group of British émigrés was the development in 1892 of a new type of instrument, more accurate than any previous. It was known as the Milne seismometer, although it incorporated some of Gray's design work as well.

Milne must have cut an incongruous figure in Japan. He was a quintessentially English character: genial, good humored, instantly recognizable by his slouch hat and Elgarian moustache. He was a keen golfer, and from his appearance one would imagine him as more at home on the golf course than in the laboratory. Nevertheless he took a keen interest in Japanese life and culture, and married a Japanese woman. He traveled around Japan with his camera, and the archive of photographs that he took provides an important window into rural Japanese life in the Meiji era.

But disaster struck—a fire that completely destroyed his home and personal laboratory. Milne decided it was time to return to Britain. In 1895 he gathered up the remnants of his papers and materials, and he and his wife set sail for England.

The location he chose was Shide, an insignificant hamlet south of Newport on the Isle of Wight, off the south coast of England. This obscure location now became the hub of international seismology. Milne threw himself into the task of fulfilling Rebeur-Paschwitz's vision. Copies of the Milne seismometer were sent to observatories around the world. Soon copies of the readings taken from these instruments, in Australia, Canada, South Africa, and many other countries, were being sent back regularly to Shide, where Milne and his assistants collated the data and published the results annually—in what came to be known as the "Shide Circulars." Additional instruments were installed at other locations in Britain, and of course Milne had his own instrument running in the converted stables of his house. He used to joke that he could tell when the local pub got its delivery of beer because the brewer's horse and cart left a characteristic signal on the seismogram. (Curiously Milne is the only seismologist to have a pub named after him—not in Shide but outside his hometown of Rochdale.)

But to make sense of all this data it was necessary first to understand what the records traced by Milne's seismometer actually meant. A seismogram (the written trace from the instrument) looks, to the untrained eye, like an irregularly wiggly line. But how could this be interpreted?

READING THE SIGNS

This was not such a hard question to answer, because theory was well ahead of observation in nineteenth-century physics. The equations for the transmission of waves in an elastic solid had been worked out, and now it was just a question of identifying the different wave types in the record of any particular seismogram.

Theory stated that there should be three principal types of shock wave. To visualize these, it is helpful to imagine a Slinky—the toy made of a single long spring that possesses a remarkable ability to "walk" down a flight of stairs. Imagine you have a Slinky stretched out across a table; someone is holding one end of it stationary, and you have the other end in your hand.

First, try moving your hand backward and forward in the same line as the Slinky. You will see a wave travel down the length of it—the rings will move forward and back while staying perfectly aligned. The rings move closer together (compression) and farther apart (dilation) as the wave moves. This is a compression wave, exactly as envisioned by

Michell back in the 1760s. In seismological parlance we call it a primary wave, or P-wave.

Now move your hand from side to side. Again a wave travels down the length of the Slinky, but it's a sinuous wave like a snake slithering along. This is a shear wave, also known as a secondary wave, or S-wave.

Last, pick the end of the Slinky off the table and make a vertical circular movement with your hand, keeping it always in the same line as the spring. This imparts a rolling wave down the length of the slinky. These are called surface waves, or Rayleigh waves, after the physicist Lord Rayleigh, who first described them. In any large earthquake the surface waves usually are the most destructive. They are also the same type of wave that you get in the sea, rolling onto a beach. Thales was not completely wrong after all.

All this was known to Rebeur-Paschwitz as he studied his first records. It stood to reason that all three wave types should be present. And since they could be expected to travel at different speeds, it followed that they would arrive at the instrument at different times, faster ones first. This was clearly going to be useful—assuming one knew the speed of the different waves, the relative difference in time of arrival for the different waves would tell you the distance of the earthquake from the instrument.

An Irishman named Robert Mallet made the first attempt to measure shock-wave speeds in a historic experiment carried out on a beach just outside Dublin in the middle of the nineteenth century.[4] This involved lighting a fuse on a barrel of gunpowder at one end of the beach, while Mallet stood at the other end with a stopwatch. When the gunpowder detonated, he had to use the stopwatch to time the interval between seeing the flash and feeling the shock wave under his feet. Considering the crudity of the experiment, the results he got were not bad. But this only gave an estimate of wave speed in beach sand. Since wave speed varies according to the sort of material through which the wave is traveling, seismologists needed to know the speed of transmission through the body of the planet. This was something that had to be worked out by degrees, gathering as many records as possible and calculating what velocities would provide the results observed.

Doing this with instruments is a lot easier than standing on a beach with a stopwatch and a barrel of gunpowder. A seismometer contains a clock, which enables the seismologist to get an exact time quite easily. With early instruments one paper chart usually represents one day's continuous recording. The trace of an earthquake is clearly seen as a series of rapid spikes interrupting the usual flat line of the unmoving pen. To mark the time, the pen would write a blip (or leave a gap) every minute, and a bigger blip every hour.

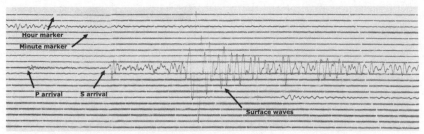

An example of a classic seismogram. This shows a Greek earthquake on October 21, 1953, recorded on a Galitzin-type seismometer at Kew, London.

When removing the paper chart from the instrument, the seismologist first checks to see if anything of interest was recorded—it might all be blank. But if he finds a back-and-forth oscillation of the pen, indicating an earthquake, he can look at the time blips on the record and determine the exact time the first earthquake wave arrived at the instrument. (Today the records are all digital. No more paper.)

The first wave is always the P-wave; the seismologist then scans the rest of the record for changes in the pattern (spikes closer together, farther apart, or suddenly getting larger) that indicate the slower waves, S- and surface, trailing in behind, and notes the time of the wave arrivals (also referred to as arrival times).

The ultimate goal of these first investigations by Milne and his contemporaries was to produce travel timetables that would tell them the expected intervals of the different wave arrivals for earthquakes at different distances.

But now some strange things appeared. Seismograms turned out to be more complex than they had expected. A lot more arrivals were popping up than just the three expected, and sometimes arrivals were missing. But the pattern was consistent, with the same anomalies occurring at the same distances.

One of the unexpected arrivals offered a clue to what was going on. This particular arrival tended to come after the P-wave, and it tended to look rather like the P-wave, only weaker. This could only mean one thing—the P-wave was splitting into two similar parts, and one was taking longer to arrive than the other. A Croatian seismologist, Andrija Mohorovičić, figured out that this was caused by a major change in density, and thus structure, about thirty kilometers down inside the earth.

Waves do odd things when changing from one medium to another. The simplest demonstration is to stick a toothbrush into a glass of water. Look at it, and the toothbrush appears to be bent exactly where it enters the water. The reason for this is that light is also a wave and is affected in two ways by the interface of the liquid water (dense) and the gaseous air (not dense at all). Some light is reflected and bounces off the surface of the water, and the rest is refracted, or bent, making it look as though the toothbrush is bent.

Mohorovičić (one of the few seismologists ever to appear on a postage stamp) deduced that something like this was happening deep inside the earth.[5] If there was a sharp change in density, some P-wave energy would bounce off this interface. Given that waves radiate out from a fault in all directions, the reflected waves (traveling down and then up again) would arrive a little later than the P-waves that took the direct route. This discontinuity is now named after Mohorovičić, usually shortened to Moho.

This was a major finding, showing that scientists could unlock the mysteries of the planet's interior. Maybe one couldn't dig deep down to look at the interior of the earth directly, but now it was possible to use earthquake waves to probe the planet's structure and determine something about what exactly was down there.

Further calculations showed the presence of another, deeper discontinuity. The pattern of missing arrivals could be explained only by the presence of liquid deep down within the earth. Since shear waves can't travel through liquid and P-waves would get bent at the boundary because of refraction, this would result in their missing certain locations on the earth's surface altogether. In contrast, surface waves, as their name suggests, travel only on the surface and would be unaffected. Hence the waves must be encountering a change from solid material to liquid inside the earth.

So in this way the interior of the earth was mapped out—a thin, relatively light crust on the outside, twenty to thirty kilometers thick, then a denser mantle going all the way down to a depth of about 2,900 kilometers, and then a liquid core in the middle (which was revealed later to have a solid "inner core" right at the center).

Now that seismologists had worked out the inner structure of the planet, all that remained to discover was what the deep-earth processes were that caused earthquakes. In Milne's lifetime, and for some decades after, this remained a puzzling mystery, solved in the end because of something utterly unrelated to earthquakes.

The Cold War.

THE SHADOW OF THE BOMB

These days we remember the 1960s as the swinging sixties, or, as Robin Williams put it, "If you remember the sixties, you weren't there." People tend not to recall so much the sense of dread that hung over the decade. With both the United States and the Soviet Union expanding their stockpiles of nuclear weapons that could wipe out humanity several times over, the constant fear was that some minor incident would be the spark that would lead to all-out nuclear exchange.

As both sides upgraded their arsenals of destruction, they needed to test new generations of weapons. But by the early 1960s radioactive fallout from repeated tests had begun to cause alarm. To try to curb this problem, the United States, Britain, and the Soviet Union signed the Limited Test Ban Treaty in 1963, prohibiting atmospheric and underwater tests of nuclear weapons. Underground testing could continue.

What was needed was a way to monitor nuclear tests, partly to police the 1963 treaty and partly, from the Americans' point of view, to keep abreast of what the Soviets were up to. Seismology offered a way of doing this.

The seismologist has always found explosions to be both a help and a hindrance. Since Mallet and his barrel of gunpowder on an Irish beach, seismologists have used controlled explosions for various experiments. But they inevitably end up recording a lot of explosions they're not interested in and that get in the way of monitoring real earthquakes. One seismological observatory in western London, for instance, remained in operation throughout World War II, and the seismograms, now carefully preserved for posterity, provide a bomb-by-bomb record of the Blitz—the German bombing raids carried out by the Luftwaffe from 1940 to 1941.[6]

A good-sized nuclear explosion, though, can be recorded at large distances. The largest of the US nuclear weapons tests at Bikini Atoll, code-named Castle Bravo, clocked in at the equivalent of fifteen megatons of TNT in 1954. Such a blast could be detected virtually anywhere on the globe.

Suddenly seismology was politically, and militarily, important.

At first it was easy to discriminate the Soviet tests from natural earthquakes. They were conducted with typical military precision, and every explosion detonated precisely on the hour. Eventually the Soviets recognized that this was a bit of a giveaway and subsequently used considerable ingenuity to try to disguise nuclear tests as natural events. They would even wait to test until a large earthquake occurred somewhere else in the world, with the idea that the shock waves from the nuclear test would be suitably masked and confused by the real earthquake waves from some far-removed location.

What developed was a race between seismologists and weapons testers, as the seismologists devised techniques for discriminating between the seismograms of nuclear explosions and those of real earthquakes, and the weapons testers looked for new ways to confuse the seismologists.

What those tasked with watching for nuclear explosions needed above all were good quality data and lots of it. In 1960 the seismic monitoring of the planet still left much to be desired. Some areas were well covered with recording stations, principally Europe and the United States, but others were a blank, and many stations were still using out-of-date equipment insufficient for the requirements of nuclear monitoring.

Improving seismic monitoring was not high on the agenda of many countries. Some stations were little more than amateur operations. But in the Cold War years of the early 1960s improvement had become a priority for the United States, and the US government was prepared to invest serious amounts of money to make it happen.

The plan was for a new global seismic monitoring network comprising 120 stations sited as evenly as possible over the whole planet, with modern seismometers at every station. It was christened the WWSSN—the Worldwide Standard Seismograph Network. By 1964 it was up and running, with stations from Alaska to the South Pole scattered across sixty countries. Every month station operators would package up a bundle of seismograms and post them to the nerve hub of the whole enterprise, in Albuquerque, New Mexico. There the data would be processed and the seismograms microfilmed, with copies of the microfilm dispatched to a few collaborating centers.

The purpose of the WWSSN was to monitor nuclear tests, but a side effect was a sudden and dramatic boost to the capability of seismologists to monitor earthquakes around the world. If 1889 marked the jump from purely descriptive reports of earthquakes to instrumental observation, 1964 marked an equally significant jump. From patchy and incomplete files of earthquakes, suddenly seismologists had available a highly complete and comprehensive view of how and where earthquakes were occurring. It was like a short-sighted person receiving a pair of spectacles for the first time. Suddenly things that once were a vague blur swim into focus. And some rather interesting patterns quickly became clear.

Geologists had long recognized that faulting could occur in three different basic ways. The first type of faulting is where the block on one side of a sloping fault line slips downward—this is known as *normal faulting*. The reverse case, where one side is pushed up over the other, is called (unsurprisingly) *reverse faulting*. You might think there is not any real difference between normal faulting and reverse faulting; in both cases one side always ends up higher than the other. The big difference is that, because such faults are always sloping at an angle, a block that moves downward actually moves away from the other block as well, whereas if one block is pushed upward along the fault line, distances across the fault are shortened. So normal faults occur when the rocks are being pulled apart, and reverse faults occur when they are being pushed together.

The third type of faulting is called *strike-slip*, or *transverse*, or *transform faulting*. Here the two sides of the fault slip past each other sideways without any vertical displacement. (It's also possible to combine an element of strike-slip with normal or reverse movement—this is called oblique faulting.)

By good fortune, just before installation of the WWSSN began, seismologists devised a technique for analyzing seismograms so that you could tell what sort of fault produced it. The difficulty with this method is that it relies on having recordings of the earthquake from every direction. This was fine if the earthquake occurred in the middle of California, where there were plenty of recording stations, but not so good if it was somewhere off in the Pacific. But the WWSSN changed all that. Now, for an earthquake almost anywhere, seismologists could get enough records to work out the type of fault movement. This meant they could also see the direction of the forces acting on the fault. And the pattern turned out to be rather interesting.

THE PIECES OF THE PUZZLE

A big clue had been staring everyone in the face since the sixteenth century, when the first moderately accurate world maps were published. Once you spot it, it's rather clear that if you could slide South America and Africa together, they would fit like a couple of pieces in a jigsaw puzzle. The fit looks too good to be purely a coincidence. The first person to draw attention to this apparent fit was the Flemish cartographer Abraham Ortelius, who published a major world atlas in 1564. But if the two continents were once joined, how did they ever separate? The only available answer seemed to be some sort of catastrophe—perhaps linked to the Great Flood of Noah's time.

And there the matter rested until the early twentieth century, when geologists started to realize that the coastlines of South America had rock formations that matched up with those in Africa. This could certainly not be a coincidence.

At this point a character named Alfred Wegener enters the story. Wegener was a German meteorologist (and amateur balloonist) who in 1912 began advocating the idea that all the continents had originally been joined together in one big landmass, which he called Pangea. At some point in the past Pangea had split up into pieces—continents—and these were slowly drifting apart.

Wegener had trouble getting his ideas accepted, in part because he was a meteorologist, and geologists were a bit sniffy about an outsider coming along and proposing a radically new hypothesis that would up-set existing theories about how the world worked.[7]

In fact, the main reason why Wegener got no traction for his continental drift theory was that while he was generally right that continents are mobile and the Americas had drifted away from Africa and Europe, and India had drifted away from Australia, and so on, he was wrong about how it all worked. He proposed that the rotation of Earth drove the process of continental drift, but a few calculations showed that this idea was unworkable. This made it easy for critics to reject his work completely.

But while Wegener and others were considering why the continents fit together, they ignored half the problem. Geologists had plenty of information about the makeup of the land but not much about the floor of the oceans that occupy more than half the planet. All that changed after World War II, when extensive surveying of the sea floor began. It turned out that the crust underlying the oceans was fundamentally different from the crust of the continents. For one thing, the crust was much thinner—a mere six to seven kilometers. And it turned out to be made of much denser material.

Meanwhile the study of the magnetic properties of rocks was about to provide more crucial evidence. It's common knowledge that if you put iron filings near a magnet, they will align themselves according to the magnet's magnetic field. The same thing happens on a grand scale with volcanic rocks. When lava erupts from a volcano as a liquid, crystals within it are free to move about and align themselves with Earth's magnetic field (assuming they are crystals of iron-related minerals). When the lava cools and solidifies into rock, these crystals are frozen in place and can't move again, even if the properties of the magnetic field around them change. In the 1950s some scientists started to study the alignment of iron-related minerals in different rocks and found some curious results. Rocks often showed alignments that were completely at odds with the local magnetic field, and that made sense only if the rocks had been formed in a completely different part of the planet. It seemed Wegener was right after all—the continents were moving.

This led to a new idea, which was being bandied about around 1960: if South America was moving away from Africa, then oceanic crust must be filling the gap between them. Studies of the alignment of magnetic minerals in the ocean crust indicated a weird pattern that further confirmed this suspicion. The whole floor of the Atlantic was striped with north-south bands. Within one band the crystals would be aligned pointing north, and the next band, pointing south. The pattern alternated, north, south, north, south quite regularly, and the bands were more or less even in width.

Such a pattern could be explained in only one way. Earth's magnetic field must flip every so often, so that the North Pole becomes south and vice versa. Also, a south-oriented band must represent ocean crust that was formed when Earth's magnetic field was in one configuration, and the north-oriented band next to it was crust that formed during the next phase. Given a constant rate of formation in the middle of the ocean, where surveys of ocean depths showed an interesting-looking north-south ridge, the observed pattern of even-width bands would be the result.

So by the time the data from the WWSSN project was starting to pour in, it was already becoming clear that the continents were indeed on the move, and the oceans were pushing them apart. The WWSSN data showed how it all fitted together.

If the continents had at one time all been one big Pangea and subsequently split, where exactly were the joins? This was now revealed in the new maps of world earthquake activity. Maps of world seismicity were not new—in fact, for the first attempt to show the global distribution of earthquakes, once again we turn to Robert Mallet, who produced a beautiful colored map in the mid-1850s that is largely accurate to this day. However, Mallet had to work with descriptions of earthquake effects, so while he could work out the major earthquake zones on land or where there were islands, he had no idea what was going on in the middle of the oceans.

The new maps, far more complete than before, demonstrated that the Mid-Atlantic Ridge was highlighted by a chain of earthquakes running along it. Furthermore they were all normal faulting earthquakes. The effect of normal faulting, remember, is that it moves blocks apart and results in the extension of the crust. So these earthquakes showed that the proponents of the notion that the sea floor was spreading were right.

But in other places different patterns emerged. In the Caribbean, for instance, along the line of the Lesser Antilles, from Antigua to Trinidad, earthquakes that happened were consistently reverse-faulting events that shorten the crust. The recorded depths of the earthquakes also appeared to get progressively deeper from east to west. So a pattern could be seen: at the latitude of the Caribbean, crust was being formed at the Mid-Atlantic Ridge, moving westward, and then was being pushed downward under something. Under what? Under another bit of crust. On the other hand farther north there was no sign that earthquakes were dipping deeper and deeper under the eastern seaboard of the United States.

But one scenario makes sense of this. After the Pangea supercontinent broke up, it split into different chunks, with the bulk of the Caribbean on a different chunk from the rest of the western North Atlantic. In one place the Atlantic crust was colliding with the Caribbean and getting pushed under it, like pushing one rug under another, whereas farther north the Atlantic crust was still attached to the North American landmass and everything from the Mid-Atlantic Ridge to Oregon was moving as a single unit.

Meanwhile another phenomenon was consistently observed in places like California. Nearly all the earthquakes were strike-slip faulting, showing that the crust was neither extending nor colliding; bits were simply moving sideways past each other.

The term *crustal plate* was introduced as a better way of describing the different units than "a chunk of crust moving about." From this we get *plate tectonics,* which is the term for the general process of plates shuffling around across the surface of the planet.

Now it was possible to start mapping the basic divisions of the earth's surface. Wegener's continental drift was really a misnomer, because it's not the continents that are moving but the plates. There might be a plate called the North American Plate, but not all of North America is on it. As Reid deduced, the coast of California is moving north relative to Napa Valley. They may be in the same state, but they are on different plates.

The boundaries of these plates are where most earthquakes occur, and one can distinguish three basic types. The first is the *constructive plate boundary,* where plates are moving apart and new crust is being formed between them. These are almost all in the oceans, but you can see such a plate boundary on land in Iceland. If you visit the site of the ancient Icelandic parliament, the Thingvellir, the path from the car park to the ancient site leads down a chasm between two rock cliffs. One cliff is North America and the other is Europe; you are literally walking along the plate boundary.

Where two plates collide, you have a *destructive plate boundary.* These come in three types, depending on what sort of crust is colliding with what. It can be ocean crust colliding with continental crust, as in the case of western South America. Here the denser oceanic crust is always forced under the lighter continental crust. This process of one plate being pushed under another is called *subduction,* and an area like the coast of Chile is known as a *subduction zone.* One sign of a subduction zone is a deep trench in the seabed that marks the start of the plate's downward descent.

Also, oceanic crust can collide with another bit of oceanic crust. One will go under the other, and the result is also a subduction zone. The Pacific has several examples of this.

Two continents can collide, in which case they crumple up like two trains in a head-on collision. The best example of this is where India has bashed into the rest of Asia—the mighty Himalayan mountain chain is the scrunched-up result.

Finally the California case is known as a *conservative plate boundary*, as crust is neither created nor destroyed.

In addition to being able to identify different plates, seismologists could work out, again from the pattern of earthquakes at the boundaries, in which direction the plates were traveling, and at what speed. It is possible to measure the distance the fault slipped as a result of each earthquake. If during a few decades a hundred earthquakes occur along the line of a plate boundary, calculating how far the plate has slipped overall during that period is a straightforward matter, given that the movement will not be completely uniform all along the boundary. Perhaps the northern part will slip a bit in one earthquake, and then five years later the next segment to the south will move in another earthquake and catch up. And so the process goes on, and all the while the bulk of the plate is moving up behind. Typical speeds are of the order of four to eight centimeters per year—which coincidentally is about the same speed at which your fingernails grow.

THE POWER SOURCE

The last piece of the puzzle is what provides the basic motive force. All the energy moving the tectonic plates around, and causing earthquakes in the process, has to come from somewhere. And here we are back in the realm of speculation rather than observation, but this speculation is based on so much data and supported by such detailed theoretical modeling that it is accepted as accurate.

The driving force behind the whole system is generally believed to be convection. If you have ever watched a saucepan of water being heated over a flame, you will know the basic process. The flame heats the water at the bottom of the pan in the middle. The warmer water rises up to the top and spreads out as more and more warm water continually rises up in a column. Then it sinks down at the edges of the pan, is sucked toward the center to replace the warm water that is rising, gets heated again, and rises to the top again. In the process the water gradually gets turned over and over.

The deeper you dig down inside Earth, the hotter it gets, right down to the molten core. The source of the heat is probably radioactive decay of minerals deep within the mantle. It was once thought that the planet started its existence as a ball of molten material spun off from the sun and that the planet's interior heat is a residual feature. Current theory states that the planet actually coalesced from dust, and the internal heat arose from pressure and radioactivity.

Although the mantle is solid enough for shear waves to travel through it, it acts like a viscous liquid that can flow. So we can imagine that what happens in the heated saucepan occurs on a vastly larger scale—hotter mantle material wells upward, spreads out, and sinks back down again. In doing so, it drags around the continents that are floating on top of it. (So Thales was not entirely wrong about the land floating—just that it is not floating on the sea.)

At the surface the system acts a bit like a conveyor belt. New crust forms mid-ocean at boundaries like the Mid-Atlantic Ridge, moves away, collides with another plate, gets pushed down beneath it, and eventually is pushed deeper down into the mantle, where it melts and disappears. Meanwhile more crust is being steadily created, and so on.

From mantle convection to plate tectonics to elastic rebound—the basic explanation of why and how earthquakes occur is all there.

Inevitably, when plate tectonics was first introduced there were doubters. Some scientists are always so committed to old explanations and old ways of thinking that they have a natural resistance to anything that upsets the old order. It's been suggested that the rate at which science advances is the rate at which old university professors die off and are replaced by younger staff who have adopted new ideas. But while there is still debate about some details of plate tectonics, the basic principles are so firmly established and fit so well with observations that it is unlikely a significant challenge will ever arise.

That said, there is still a lot to learn. It's possible to find a map in a textbook or on an Internet page that purports to show the divisions of the world into tectonic plates with nice neat lines slicing off Europe from Africa and so on. In truth the situation is much messier. For instance, after the Japanese earthquake of March 2011 journalists had some difficulty explaining the tectonic situation to readers. Yes, eastern Japan is a subduction zone. Yes, the Pacific Plate is moving westward and is being subducted under the Japanese mainland. And the Japanese mainland is on which plate, exactly?

The obvious answer is that Japan is part of Asia, so it must be part of the Eurasian Plate. But one could also find press releases stating that Japan is on the North American Plate. On the face of it this seems odd, since Japan is on the other side of the Pacific, but in fact the North American Plate extends across the Bering Strait into Siberia. Again, continents and plates are not the same at all. The final answer is that no one is entirely sure, and geologists are divided into four camps, arguing for the Eurasian Plate, the North American Plate (extended across the Bering Straits from Alaska), the Okhotsk Plate (a much smaller plate encompassing chiefly eastern Siberia), or the Honshu Microplate (a microplate is a small crustal plate; this one is supposed to have broken off from the Okhotsk Plate). Three of these four possibilities propose that Tokyo is on a different crustal plate from Western Japan, with a plate boundary running north-south across the Japanese mainland.

But if this geological division is a boundary between different plates, one would expect to see earthquakes along it. Which one doesn't. So there is plenty of scope for discussion—and disagreement.

Which plate is Greece on? Textbook maps show clearly that it's part of the Eurasian Plate. But recent thinking suggests that so many bits of plate have been mashed up together in much of the Balkans that trying to identify it as one plate or another no longer makes much sense—it's where two plates merge.

It's a mistake to take simple analogies too far. One shouldn't imagine that each tectonic plate is like a vast piece of crockery, with sharp, well-defined edges and rigid throughout. This is one reason why although the vast majority of earthquakes occur at plate boundaries, not all do.

Earthquakes away from plate boundaries are called intraplate earthquakes. They occur precisely because crustal plates are not uniform, rigid slabs of rock. Much continental crust has been around a long time and has been thoroughly broken, bent, and scrunched around by past geological disturbances over the millennia. So the slabs of continental crust that are being dragged around by tectonic forces are more like complex assemblages of rock stuck together, and it shouldn't be too surprising if sometimes there's a bit of slippage on any one of the numerous old faults that criss-cross anywhere you care to mention.

While intraplate earthquakes are generally much less frequent and smaller than those on plate margins, they can be important nonetheless—many major cities are situated in intraplate areas, and a direct hit on New York City or London from even a modest earthquake could be significant. Even away from earthquake zones, you are never completely safe.

4

TRACKING THE UNSEEN

KANT WAS NOT ALONE IN LINKING EARTHQUAKES AND VOL-
canoes. And not without reason. Many of the world's earthquake-prone
regions are also volcanically active. Take, for example, Japan's Mount
Fuji or the line of volcanoes extending from Mexico down the spine
of Central America and all the way down the Andes to southernmost
Chile. It's not just coincidence—volcanic chains tend to run parallel to
subduction zones, where the destruction of oceanic plate material, as
it's pushed down into the mantle, releases water vapor that ultimately
works its way upward and triggers volcanic activity. Subduction zones
are also where the greatest earthquakes occur.

But this isn't the only link between earthquakes and volcanoes. The
movement of magma beneath a volcano before and during an eruption
is not a smooth flow, and it causes many minor earthquakes. Tracking
these earthquakes is one way in which volcanologists can tell that an
eruption may be imminent. One branch of seismology—volcano seismology—is devoted to the study of volcanic earthquakes.

But the day-to-day practices of studying volcanoes and studying
earthquakes tend to be rather different. This comes out in one curious
statistic—it is more dangerous to be a volcanologist than to be a seismologist. If you ask volcanologists to name someone killed while making
observations of an erupting volcano, they will have no difficulty—and

could even cite the example of the Roman scientist Pliny, who died because he got too close to Vesuvius during the great eruption of 79 A.D. In contrast most seismologists would look blank if asked to name a seismologist killed by an earthquake. I only know of one case, not well known, at Tangshan in 1976, when a party of seismologists was breaking their journey back to Beijing and just happened to be stopping over the night of the earthquake that leveled the city.

It comes down to duration. An earthquake, whether it is knocking over wine jars in ancient Greece, rattling Oxford windows in 1666, or shaking up San Francisco in 1906, is over quickly—a couple of minutes at most. A volcanic eruption can last for weeks. So if a volcanologist hears that Vesuvius is erupting, she has plenty of time to pack her bags, book a plane ticket, and arrive in Naples while Vesuvius is still erupting. The seismologist arriving after an earthquake can survey the ruins, record the aftershocks, and measure the surface break of the fault, if there is one, but the main event has been and gone.

Some seismologists do work in this way. But there is a limit to how many people it is actually useful to have on the ground making investigations, especially since most countries will have a local team responsible for undertaking such studies. This is especially true after a major earthquake disaster; the highest priority must be given to search-and-rescue operations, and it's not always the best idea to have seismologists swanning around looking at the science and getting under the feet of those who are trying to dig survivors out of the wreckage.

Seismologists do travel a lot, but it's mostly to attend conferences or workshop meetings, to see other seismologists and set up collaborations. Seismology has always been an international subject. Earthquakes don't respect national borders. Seismologists can do far more through cooperation with colleagues in other countries than on their own.

But another reason for not traveling to look at earthquakes is that, to a large extent, the seismologist doesn't have to: the earthquake does the traveling. In my case I have a fine view from the windows of my top-floor office: looking west I can see over the golf course next door to the observatory on top of the hill. This splendid Victorian edifice, with its dark stonework and two massive copper towers housing the main telescopes, has an association with seismology that goes right back to Milne's time. The seismometer vault is still in operation today; if a large earthquake strikes Japan, the instruments in that vault will start twitching about eighteen minutes later.

If I want to look at the data, I don't even have to walk to the observatory (just as well; the hill is rather steep). Thanks to modern communications technology, I can stay in my office, call up the seismometers from my computer, and see the seismogram plotted on my screen. In fact I can call up any instrument on the national network if I choose. Or I can look at the essential readings (if not the original waveforms) from monitoring stations all over the world.

DRAGONS AND SOUP

We do know the name of the first man who had the idea of detecting earthquakes at a distance. He long predated John Milne and his colleagues in Japan at the end of the nineteenth century. We have to go back to 132 A.D.—in China. The inventor in question was named Zhang Heng (the spelling varies according to the transliteration system used). He was a man of many talents: an astronomer and mathematician as well as inventor, and, as was expected of any man of learning in the Eastern Han dynasty, a cultured poet.

He is remembered most today for something he called *houfeng didong yi*—a device for inquiring into wind and movements of the earth, which he presented to the court and installed at the imperial palace. The instrument consisted of a large jar with eight dragons' heads arranged in a circle near the top. Each dragon held a ball in its mouth. At the base of the jar were eight frogs with gaping mouths. The idea was that if an earthquake occurred in some remote part of the emperor's domain, the distant vibration would be enough to cause one of the dragons to drop its ball—*ka-ching!*—into the frog's mouth below. By seeing which dragon dropped its ball, you could infer the direction of the earthquake. Reportedly, it worked. Exactly how, no one now is sure, but it is speculated that a weighted pendulum suspended inside the jar could swing in any direction, with some sort of arrangement of rods attached so that if it did swing, one ball would get a knock that would dislodge it. A visitor to China today may encounter nonfunctioning brass reproductions of the device for sale; they make nice seismological souvenirs.

Various attempts have been made to reconstruct this instrument, with varying degrees of success, the first being in China about five hundred years after Zhang Heng's time, and later in modern times. The most recent reconstruction in 2005 by members of the Chinese Academy of Science in Zhengzhou, Henan province, is claimed to be the most accurate, and most effective, yet.[1]

It would be a long time before the West began to catch up. One story has it that the practice in Italy was to set out bowls of thick soup. If you inspect the bowl in the morning and find it has slopped up one side, you know an earthquake has occurred during the night.

But Zhang Heng seems to have had the right idea back in the second century—a pendulum offers the possibility to detect even faint earthquake motion. People undoubtedly noticed hundreds of years ago that, for example, hanging light fittings gently swing at the time of a distant earthquake. By the 1700s Italian scientists were trying to use simple pendulums specifically for the observation of earthquakes.[2]

EARTHQUAKE SWARMS

Earthquakes generally don't come singly. After a fault has violently wrenched itself into a new configuration, after decades of being locked rigid by friction, the new position usually is not stable. Until it gets

completely locked into a new position, it goes through a sort of settling-down period when parts of the fault (and sometimes neighboring faults as well) adjust, which takes the form of a series of smaller
earthquakes we call aftershocks. And sometimes there is a series of little
premonitory earthquakes called foreshocks (the big one is called the
main shock).

But sometimes it doesn't work that way. Sometimes you get a fault
that, for one reason or another, doesn't lock again easily. So instead
of a few foreshocks (or none at all, which is common), a main shock,
and then a series of aftershocks that trail off in the next few weeks or
months, you get a sequence of earthquakes, mostly small, that just go
on and on for years, sometimes dozens per day. This is called an earthquake swarm, and some places are particularly susceptible to them.

One such example is the Scottish village of Comrie, in Perthshire.
Comrie is one of the most attractive places you could imagine. It lies
on the banks of the river Earn, which idles through a broad flat valley
called Strathearn—*strath* is the Scots word for a wide valley. Strathearn
is a pleasant landscape of deciduous trees and farmland. But just north
of the village itself is a major geological boundary, where the Highlands
of Scotland begin and rugged wild hillsides of heather and rock rise up
suddenly in contrast to the domesticated valley landscape.

Comrie developed a reputation as the "shaky town" in 1795 with a
series of small earthquakes that lasted for six years. The local minister,
Reverend Samuel Gillfillan was even awarded the title "Secretary to the
Earthquakes" for his diary recording the events. But after rising to a cli-
max in 1801, the shocks died away, and peace was restored to the valley.

Until 1839, when the shaking started again. In October of that
year a short sequence of shocks rose quickly in intensity and suddenly
stopped. And then it started raining. It rained and rained, as heavily
as anyone could remember. And then, on the night of October 23, the
strongest earthquake ever felt in Comrie sent the unfortunate villagers fleeing in panic from their houses into the downpour. It was late

in the night before the drenched and miserable inhabitants could be persuaded to enter the church for an impromptu service to pray for deliverance.

In world terms the earthquake of October 23, 1839, was a minor affair, and damage was restricted to a few toppled chimney pots. But the swarm had now really settled in, and during the ensuing weeks earthquakes were felt in the hundreds.

Farther afield, heads were turning in the direction of the little village. Earthquakes pass quickly, and normally a scientist could expect to wait a long time before experiencing another one, but this was something different. Earthquakes were happening so often in Comrie that you could mount an experiment and be pretty sure that an earthquake would come along while you were running it. The village had become a natural laboratory for studying earthquakes; this was a chance not to be missed.[3]

So at its 1840 meeting the British Association for the Advancement of Science formed a special committee to investigate the earthquakes and see what could be learned. Leading the team was a certain David Milne.

It is a strange coincidence that two major pioneers of earthquake studies in Britain in the nineteenth century should have had the same surname; they were not in any way related and never met. David, the earlier Milne, had a lot in common with Charles Lyell, the geologist (see Chapter 2). David Milne was an Edinburgh lawyer who found law not to his taste and geology a more congenial subject. He married into the wealthy Home family (pronounced *Hume*) and later changed his name to David Milne-Home. But in 1840, while still supposedly running a legal practice, he decided that earthquakes would be an interesting study and neglected his law cases in favor of taking on the task of presiding over the earthquake investigation committee.

The most important member of Milne's committee, other than Milne himself, was James Forbes, professor of Natural Philosophy at

Edinburgh University. Forbes was given the task of designing an instrument to record and, they hoped, measure the earthquakes. His design was based on the pendulum but with a twist—the pendulum was to be upside-down, hinged at the base, instead of suspended from the top.[4] Since an inverted pendulum is inherently less stable than a common pendulum (the type suspended from the top, such as is found in a grandfather clock), the instrument had the added advantage of being more sensitive. (A similar instrument would be invented independently in Prague a few years later.)

The instrument consisted of a round wooden base with a metal rod fixed in the center such that it was free to swivel. Near the top of the rod was a metal weight, and right at the top was a pencil. Fixed above the pencil was a shallow dome that could be lined with paper. The idea was that when the pendulum moved in response to a shock, the pencil would move about, writing a record of its movement on the paper.

Forbes built several of these instruments in different sizes. The biggest was reputedly three meters tall, but most were small and could easily be carried around. Forbes had invented the first portable earthquake-recording device, but what to call it? Milne had a suggestion. He turned to Greek (the mark of a cultured man in those days) and put together *seismos* (earthquake) and *metron* (measure) and came up with *seismometer*—a measurer of earthquakes. This was the first time any *seismo* word ever appeared in English. *Seismology, seismologist,* and all the other similar words are all back-formations from Milne's coinage.[5]

Armed with Forbes's invention, Milne threw himself into the Comrie investigation. He set about collecting information from all over Scotland to ascertain how widely the stronger shocks had been felt. Meanwhile a bunch of the new seismometers were installed in and around the village, overseen by Peter Macfarlane, the local postmaster and storekeeper in Comrie. Macfarlane also kept a register of all the shocks that were being felt day in, day out, in the village.

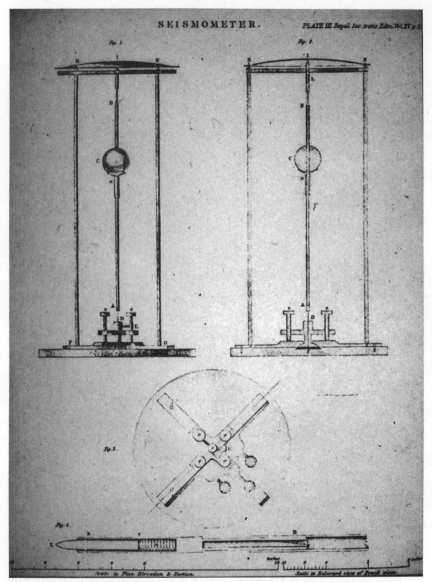

The original engraving of James Forbes's "seismometer" from 1841. David Milne's coin-age of the word seismometer set the pattern for words like seismology and seismologist.

Milne's final report of the investigation makes interesting reading today.[6] He made no epochal discoveries, and Forbes's instruments were still insufficiently sensitive (the problem was friction of the pencil on the paper) and recorded little. But Milne had some interesting insights, including a means of estimating the depth of an earthquake source.

He was pretty sure that earthquakes and faults were connected, but he couldn't be sure exactly what the connection was.

As for Forbes's instruments, one survives today in the Royal Scottish Museum in Edinburgh but sadly lacks the dome part into which the paper would be tucked. It is rumored that at least one other exists in private hands. None of the few records made by the instruments still exist.

But there is another curious little note in the committee's proceedings. Apparently, Macfarlane played around with some other designs of instruments. He built a little observatory in a cupola that stuck out from the roof of his house (which was above his shop on the main street of the village). The building is still there today, but the cupola was destroyed long ago in a fire. One of these instruments is described as being a horizontal pendulum. Who made it is not recorded. But it may well have been the first use of a horizontal pendulum in the study of earthquakes.[7]

PITS AND PENDULUMS

The horizontal beam pendulum would prove to be the fundamental design for earthquake-measuring instruments for the rest of the nineteenth century, up to and including the period in which John Milne was active. The idea is simple: take an upright pillar and fasten a hinged beam to the base. The end of the beam farthest from the pillar is fastened to the top of the pillar by a wire. The beam can swing to and fro. At the very simplest, if you attach a pen to the end of the beam and put a rotating drum covered with paper under the pen, the pen will record any swings the beam might make. Since pen and paper create friction, seismologists eventually realized it was preferable to use a narrow beam of light in place of a pencil, and to cover the drum with photographic paper.

The principle by which this arrangement works is not obvious at first. You might think that the earthquake waves were nudging the beam and causing it to swing. But actually the ground is moving, not the pen. As the ground trembles, the movement is not directly imparted to the pendulum beam, because of inertia and because the beam is not rigidly attached to the pillar. The ground moves, and with it moves the pillar, the recording drum, and the whole room—everything except the pendulum beam. Because the beam is quite long, a small relative movement close to where it is hinged is transformed into a large movement at the far end, thus magnifying the small ground movement into something large enough to be clearly visible.

A bit of terminology is helpful here. Although the word *seismometer* was coined specifically for Forbes's instrument, a different word is used today. Forbes's device, and for that matter Zhang Heng's as well, would now be called a *seismoscope*. A seismoscope makes some sort of record if an earthquake occurs, but it doesn't tell you when it occurred. If Zhang Heng got up in the morning, went to look at his jar, and found a ball had dropped, he could tell an earthquake had happened in the night but not exactly when. The bowl of soup yields the same sort of information.

A seismograph, on the other hand, has a clock attached to it, so you can see not only that an earthquake has occurred but precisely when. The first seismographs were invented in Italy in the 1850s—credit is given to Luigi Palmieri, who lived and worked in Naples, eventually becoming director of the Vesuvius volcano observatory.[8]

Today, the word *seismometer* refers to an instrument that not only records the time but makes an accurate record of how much the earth moved. These really started with John Milne, and almost any modern instrument will technically be a seismometer. Some early designs maximized inertia by increasing the weight of the pendulum. At the extreme, one design had a pendulum that incorporated a huge iron bucket full of boulders. I saw one once in an observatory in Strasbourg; in the semidarkness of the vault this huge construction of girders and stones seemed to grow out of the corner of the room, lurking ominously like something out of a science fiction story.

Building instruments on that scale was only ever practical if only a few were needed, and you could more or less build an observatory around them. Something more portable was needed if, like John Milne, you wanted to send instruments off to other observatories around the world.

In modern seismometry portability is even more important. Modern seismology requires detection of small earthquakes as well as large ones—for instance, if one wants to study the seismicity of a not particularly active area such as eastern North America or Western Europe, a few observatories are not going to be enough. To capture small earthquakes you need to have an instrument reasonably close by; to monitor an entire country you need a network of instruments spread evenly around.

So it's highly useful to have an instrument small enough and sufficiently self-contained that you can deploy it somewhere other than in a staffed observatory. You want to be able to dump it somewhere in the countryside and let it run on its own.

A modern seismometer is a little metal cylinder about the size of a soup can. It still works on the pendulum principle, but the pendulum is a cylindrical weight inside the can, attached to the casing by means of a spring. Come an earthquake, the ground moves, and the casing moves with the ground, but because of inertia the weight, which is insulated from the shaking by the spring, is not directly affected. However, inside the casing some electromagnets are fixed to the sides; these stop the weight from moving relative to the casing. This way, when the ground moves, the whole instrument, including the pendulum, moves with it.

For example, if the ground moves up, the casing moves up, and the weight is pulled up by the magnets so that it stays in place relative to the magnets. Because these are electromagnets, the force they apply to hold

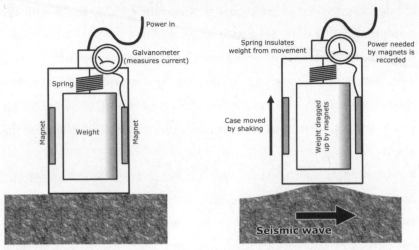

This gives a very simplified view of how a modern seismometer works. Unlike the observatory instruments of the early twentieth century, these instruments are small and easily portable.

the weight in place is supplied by electrical current. The instrument records how much current the electromagnets use, and this is something that can be converted into a record of how much force was needed to stop the weight from bouncing around, creating a record of how much the ground was moving.

Today this information is recorded digitally—no need for drums wrapped in photographic paper. The digital signal can be transmitted by radio or over telephone lines; the seismologist need not visit the instrument every day to change the paper.

So a key task that a modern seismologist has to consider is installing and running a network. An even distribution of instruments across an area is ideal, but other factors are important, too. Quiet locations away from human activity are preferable. There's not much point in putting an instrument close to a major road, where it would pick up vibrations from every vehicle that passed. A better signal will also be achieved if the instrument is on rock, which makes it difficult to find good sites in lowland valleys of thick alluvium. Nice rocky hillsides as far from human settlement as possible make excellent sites.

It's also best to avoid the coast if possible—waves beating on the rocks will be picked up even significant distances away. Also, forests are bad sites—when trees blow in the wind, their roots move in the soil, and this results in a messy, noisy record. Strange as it may seem, seismograms to some extent preserve a record of the weather. The earth is never completely still. You might think a seismogram recorded in the absence of any earthquake or any other bump or bang might just be a straight line. In fact seismograms always show slight wiggles, called noise, or microseisms. These are just the cumulative effect of every possible minor disturbance. On a fine day a seismogram recorded at a good site will show such little wiggles, but they will be small. On a windy day they'll be rather larger, and on a really stormy day the record will be such a mess of wiggles it can sometimes be hard to spot the actual earthquakes.

Once a site is selected and the landowner has granted permission, it's time to install the instrument. This involves digging a little pit down to rock, putting down a layer of cement to get a good surface, and then installing a box over it to provide protection. The instrument sits in the box, using its aerial to send the signals to home base, powered by a solar panel on the aerial mast. The instrument can now be left alone on its hillside to do its work and transmit its data; unless some malfunction occurs, one might not need to return to the site for years.

FEELING THE PULSE OF THE PLANET

Meanwhile back at base the analyst is collecting and processing the data from stations across the network.

John Milne could sit in his observatory in Shide and collect earthquake waves arriving at his instruments; he could also collect readings from other stations scattered around the world as they arrived by mail. The situation is not so much different today, except that the electronic transmission of data is so much faster, and there are thousands of seismometers all over the world, instead of the small handful in Milne's day.

I rather like the idea of the job of an old-fashioned seismologist. I would quite fancy running a little observatory on a tropical island somewhere, where a seismologist's duty every day would be to saunter down the path between the palm trees after breakfast, over to the instrument vault. Then one would change the paper on the seismometers and develop the records in the darkroom. Then, carry the fresh seismograms back up to one's office, brew up a fresh pot of tea, and sit down to examine the records for any earthquakes. With a ruler and pencil, and sometimes a large magnifying lens, one would work through all the interesting waves detected and jot down the arrival times in a big ledger. Then, at the end of the month, transcribe all the readings, stuff the paper into a big envelope, and walk leisurely down to the harbor to mail the data off—if not to Milne himself, to one of his successors.

That was more or less the life of an observational seismologist, and I knew people who worked like that as recently as the 1980s—or even later in the developing world. Today the methods and the equipment are vastly different than in Milne's time. Gone are the bulky instruments requiring a specially constructed vault in the observatory precincts, replaced by those little silver cans buried in pits on remote hillsides.

Most countries operate some sort of national network for monitoring earthquakes. It's usually run by a national institute that may also be part of the country's geological survey. Sometimes there is a single national network, as in the United Kingdom. Sometimes a country will arrange for monitoring regionally, as in Germany or the United States (California has separate networks for north and south).

Since it is no longer necessary to tend every instrument daily, it's quite possible to run an extensive network reasonably inexpensively. Scores of instruments requiring little maintenance can be placed up and down the country, all linked by radio or telephone lines to the central office, where the data can be called up by computer on a regular basis.

And of course the ruler and pencil are long gone. The seismologist whose job it is to make the readings brings the seismogram up on a computer screen, zooms and pans the record to examine it carefully (no more need to use a big magnifying glass to scrutinize a faint record), and records where the key wave arrivals are with clicks of the mouse. No more need to count the ticks to get the precise time information—the software calculates that for each mouse click and automatically stores it.

In fact, up to a point the seismologist can leave the task entirely to software. Some programs can continually analyze the signals coming in from seismometers, detect earthquake waves from random noise, distinguish one wave from another, and even make an automatic assessment of the earthquake location. Such automatic systems are useful for giving a quick alert that something needs attention, but one cannot entirely trust a computer program to make the right decision. On one memorable occasion I got a sarcastic email from a colleague in eastern Europe asking if I needed help extricating myself from the rubble of my office. Someone moving about in the vault up the hill had accidentally kicked the stone block a seismometer was sitting on. This had been picked up by an automatic processing program that had decided this was a substantial local earthquake and inserted the information on an international log of recent earthquakes, which is where my friend spotted it before it could be removed.

Such a mistake should not have happened then and could not happen now. A strong signal on only one instrument can't be an earthquake—it would be detected on several stations at widely different locations if it were. Having a large network with many stations means that a lot of data can be collected, and instruments close together are more likely to detect even small earthquakes. They give a better overview of what the earth is doing at any point in the national territory.

But modern seismometers are sensitive instruments. John Milne could detect the brewer's heavy cart going by his house; a modern instrument can detect lighter road traffic and at considerable distances—as well as pick up vibrations caused by trees blowing in the wind, waves pounding on the shore, or the occasional sheep wandering past, and automated systems have to be able to know which signals to ignore.

But even the smartest computer programs are not as good as a trained human at analyzing a seismogram. I've known seismologists who were so experienced at analyzing records from their instruments that they needed just one quick glance at a seismogram to tell you where the earthquake had occurred. So even if today the first alert is a bit of automated computer analysis, the seismologist on duty will be sure to review it and make a definitive judgment on the precise readings.

And then it's time to share the data. This is no longer done by typing up the record of the month's activity and mailing it off to Shide. Transmission is done rapidly by computer, and the data may be uploaded onto the institute's website at the same time.

The first destination is the US Geological Survey in Golden, Colorado. This is the principle clearinghouse for seismic data in the short term. The USGS collates data from all over the world as they come in, calculates the earthquake parameters, and publishes the results. Once upon a time these were printed and sent out weekly; now they go straight to the web page of the USGS's National Earthquake Information Center, at http://earthquake.usgs.gov/regional/neic/. As more data are received, the parameters are gradually refined and updated. For any large earthquake of international significance, figures from the USGS are what you are likely to hear cited in the media. For a smaller event of interest chiefly to the country in which it occurred, figures released by the national agency will be more accurate.

The European-Mediterranean Seismological Centre, based just outside Paris, performs a similar function for earthquakes in Europe and the wider Mediterranean region. Both organizations have websites with constantly updating maps (see www.emsc-csem.org/); you can leave the page running in your web browser and see a new dot when a new earthquake happens somewhere. It would have seemed unbelievable to have such information available to a seismologist in Milne's day. You can now view the USGS data in Google Earth as well, and have them continually updated. This is especially useful in cases such as the Tohoku earthquake in 2011, when it was necessary to try to keep abreast of a rapidly changing aftershock situation.

But the last clearinghouse and ultimate authority is, appropriately enough, the descendant of John Milne's network—the International Seismological Centre (www.isc.ac.uk/). Milne died in 1913, and for a brief time one of his assistants attempted to carry on the work from Shide. After the First World War the task of coordinating the worldwide collection of data moved to Oxford, and the enterprise was rechristened the International Seismological Summary. In the 1960s it was renamed the International Seismological Centre (ISC) and moved first to Edinburgh and then back to the south of England, to Thatcham in Berkshire, a little west of London, where it operates out of a huge shedlike building in a small industrial estate. The ISC's mission has always been to take more time, collect the most complete data, and then publish the most definitive account possible for each earthquake.

It's a sensible division. The USGS publishes the first timely version of an earthquake's parameters, which satisfies the need of the media and the public to have some information as fast as possible. Then, sometime later, the ISC's considered verdict appears, and it will be the one used in subsequent scientific studies. And all of it is made possible only by international cooperation on a global scale.

It was recognized early on that international cooperation was essential for seismology to progress. Some fine group photographs survive of the international meetings in the early years of the twentieth century—rows of whiskery faces and hats. Traveling to an international

This group photo was taken during the first International Seismological Conference, held at Strasbourg (then Strassburg, Germany) in 1901. (Photo courtesy of Johannes Schweitzer.)

meeting in those days was rather more of an effort, involving trains and boats—no driving to the airport for a low-cost flight back then.

After World War Two a great fear in Europe was that the seismological community would be split in two by the iron curtain, and a resolute step was taken to prevent that from happening. The result was the founding of the European Seismological Commission (ESC) in 1946 (see www.esc-web.org/). The aim of the ESC was to provide a forum for seismologists from West and East to meet and exchange ideas. Every two years it would hold a General Assembly, a week-long meeting, not just for talks and lectures, but also for forming working collaborations on specific tasks. One rule was that meetings must alternate sites. If one meeting was held in a Western country, the next had to be in an Eastern country, and so on. This was a brilliant move that kept seismologists working together and talking together despite differences in ideological systems. More than sixty years later the ESC is still flourishing, though with the end of the Cold War, the need to switch venues has been shelved.

BUMPS, BANGS, AND BELLS

People become seismologists in different ways. Perhaps the most common is through geophysics, or the study of physics as it applies to planetary bodies (not just Earth). However, I know plenty of people who fell into seismology by way of other studies. Geology is an obvious one. It's a logical progression for a geologist who is interested in rocks and faulting to become interested in earthquakes. Another common background is engineering. Engineers have to work out ways to design buildings that are safe against earthquakes, and sometimes they simply slip over, drop the engineering, and become seismologists. Earthquake engineering is something that I will discuss in much more detail in Chapter 10.

Electronics is another entry point. The work of designing seismometers continues; people are always looking to see if existing models can be improved. It's not likely to be a geologist or a geophysicist who works on new instruments—such a task requires specialized expertise in electronics. Mathematics is another subject that comes to mind: seismology offers plenty of scope for mathematicians to try their hand at applying new statistical techniques to earthquake data. And that is still not all. You might think of history as an academic subject with no practical applications and certainly not something to be pursued with a view to the advancement of modern science. Yet trained historians and seismologists today are working side by side on the record of past earthquakes with the aim of better understanding where and how often earthquakes have occurred.

Seismology also has specialized fields within it—I mentioned volcano seismology at the beginning of this chapter, and explosion seismologists, whose work I introduced in Chapter 2 detect nuclear explosions. The field of *forensic seismology* is also encountered in the study of nuclear explosions, and it is a sort of detective work. It's part of a wider issue in everyday seismology—distinguishing between real earthquakes and other bumps and bangs that appear on the seismographic record. After all, a seismometer is simply recording vibrations in the earth, and an earthquake is not the only possible source.

Very local causes—someone kicking the box, a truck going past—are easy to spot with modern seismological networks, as the event will register on only one instrument. These types of things don't even have the same wiggly shape as an earthquake. One quite quickly acquires the knack of spotting obvious nonearthquakes in the record.

But some things can be detected on several distant stations, look like earthquakes, and yet are not—for example, sonic booms. If a jet aircraft passes over your house at Mach 2, you might well think an earthquake has occurred, as the walls and windows shake. It looks pretty much like an earthquake on a seismogram as well, but there is one key giveaway. An earthquake wave travels at the speed of sound in rock. A shock wave from a sonic boom travels at the speed of sound in air, which is slower. So you just have to see how quickly the wave passed from one recording station to the next. If it's slow, it's a sonic boom.

When tracking explosions, sometimes the key is that you already know it's there. If a large explosion happens as a result of an accident, say, a dynamite factory blows up in the night, it will be all over the news the next morning, and it will be quite obvious that you have indeed recorded it at the reported time. Quarry blasts are not so newsworthy, so occasionally one has to think a bit about these. If a signal corresponds to something shallow that happened during the working day, in an area where quarries are known to exist, it's probably a quarry blast. If it's five kilometers deep and at night, it certainly isn't.

Another way to be sure a signal is an earthquake or an explosion relates to the physical properties of earthquake faulting. It's easiest to imagine in terms of strike-slip faulting. Suppose you have a vertical fault running east-west, and suppose you have seismometers to the

northeast, southeast, southwest, and northwest. Now imagine an earthquake occurs so that the northern side of the fault jerks eastward and the
south side jerks to the west. What will happen at the four instruments?
On the north side the ground is moving toward the instrument at the
northeast quadrant and away from the one in the northwest quadrant.
On the other side the reverse happens—the ground twitches westward,
toward the southwestern instrument and away from the southeast.

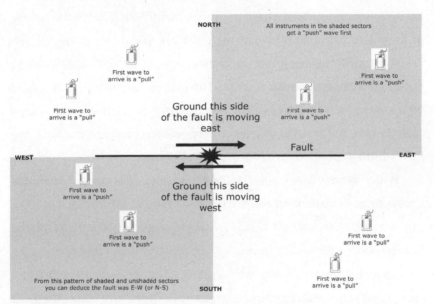

*This diagram demonstrates how the first wave to arrive at seismometers arranged around
a fault are different in each quadrant; in this case the fault is a vertical strike-slip fault.
With normal or reverse faulting the pattern is more complex. This type of analysis is a
powerful tool for recognizing what type of fault has produced any earthquake.*

Although all four instruments may get plenty of shaking, the first
wave that arrives is critical—it will be an "up" wiggle if the ground is
moving toward the station and a "down" wiggle if the ground is moving away. The first wave on the record will go up on the northeast and
southwest records and down on the northwest and southeast records.
The seismologist who sees this pattern can work out something about
what the fault was like. Unfortunately we always get two possible results

at right angles to each other: a north-south fault and an east-west one will produce the same pattern. But it's enough for the seismologist to be able to say, "This was an earthquake on a strike-slip fault running either north-south or east-west." (And this was the sort of analysis that, repeated over thousands of earthquakes, enabled the pattern of plate motions to be worked out.)

But an explosion throws the ground up in every direction equally. So recordings from different stations show every single first wave in the up direction, no matter where the instrument is relative to the source of the waves. And that is usually the giveaway that it was an explosion.

If an earthquake happens underwater, something else happens as well. The water around the blast starts to vibrate almost like a bell, and on a seismogram this has a characteristic shape of relatively widely separated peaks and troughs. A trained seismologist can glance at that and say, "Mmm—underwater explosion."

Which is why it was pointless to try to conceal the true reason for the sinking of the Russian submarine K–141 *Kursk* on August 12, 2000. The sub was lost with all hands in the Barents Sea; the cause of the sinking was an explosion in one of the torpedo tubes that triggered a second and larger explosion that doomed the sub. At first the Russian authorities were loath to admit that any error or failure had caused the loss of one of their finest submarines, and they attempted to blame the loss on collision with another vessel, perhaps a NATO spy ship operating illegally in Russian waters.

The truth could not be hidden. Seismometers in northern Norway had recorded both explosions, and the characteristic bell-like seismic signal was easily recognized.

TIME AND PLACE AND—

For a seismologist in Milne's day, the routine I described earlier was largely concerned with measuring the time at which waves arrived at the recording station. From those recorded times, collected from a large number of stations, seismologists could trace the waves to the source. Today, if the speed and timing of the waves indicate that the earthquake originated three thousand kilometers from Washington, DC, two thousand kilometers from Houston, and five thousand kilometers from Seattle, a little work with a globe will reveal that the origin is somewhere around Honduras, and a computer program will calculate the location even more precisely.

Location is a commonplace English word that is nevertheless used routinely by seismologists when talking about earthquake data. But there is, of course, a technical term as well, albeit one that is much abused in everyday journalism. That term is *epicenter.*

To understand it properly, think again about the actual physical process involved in an earthquake. The 1906 San Francisco event will do as an example. The San Andreas Fault is a long fracture in Earth's crust that runs the length of the state, and not just at the surface, but extending down maybe twenty kilometers. The western side is being dragged northward relative to the eastern side, and just before April 17, 1906, the rocks on both sides of the fault were straining at the limit of what they could bear. The fault was still held in place by friction, but that was about to give.

The fault breaks—somewhere. One little bit of rock snaps first. A small section is released; it moves, it jerks, it kicks the next bit along, and the whole thing starts to unravel. The tearing spreads rapidly up the fault line for hundreds of kilometers.

Energy is released wherever the fault breaks, so do not be misled by newspaper diagrams that show an earthquake as concentric circles on a map. That one little point where the break begins is where the first seismic waves start from. It's interesting to seismologists for that reason, but it's not always the place where the most destruction occurs—that could be farther along the fault.

Also, it's not the epicenter. The origin point, where the whole earthquake starts, is called the hypocenter. The prefix *hypo* comes from the Greek for *under;* the hypocenter is the "undercenter," and you can't really plot it on a map, because it has depth. It's also called the *focus* of the earthquake, and its depth is typically five to fifteen kilometers down and can be considerably deeper. It's important to think three-dimensionally; a fault is not just a line on a map, it's a plane and may be dipping (i.e., going down at an angle rather than being vertical) as well.

The epicenter is the point that you can put on a map. It is defined as the point on the earth's surface that is directly over the hypocenter. If a fault is gently dipping, it's possible that the epicenter may not even be close to where the fault is plotted on a geological map. Suppose, for example, a fault runs north-south but slopes down to the west at 45 degrees. If the earthquake starts at a point on the fault that is ten kilometers down, that point is also ten kilometers west of where the fault is at the surface. So comparing epicenters with faults on a map is not straightforward.

So next time you read that "Milan is the epicenter of the fashion industry" or some such expression, remember that it means that Milan is the point on the surface directly above the hypocenter of the fashion industry. Not such a good metaphor, is it?

If you look at early-twentieth-century earthquake bulletins such as were published by Milne and his successors, you will see for each earthquake the date, time, epicenter as latitude and longitude, and maybe depth (which is often omitted, as it can be quite hard to calculate depth accurately). Time and place—isn't something missing?

The alert reader may have noticed something slightly odd in these first four chapters. Whenever I've mentioned a specific earthquake, I've been rather vague and called it small or great. Usually, whenever you read anything about any earthquake, you will see a numerical expression of size introduced as a matter of course. So far I've deliberately left these off. There is a reason for that.

Let me explain.

5

HOW BIG? HOW STRONG?

THE ONE THING THAT MOST PEOPLE THINK THEY KNOW about earthquakes is wrong. Whenever an earthquake is mentioned, the first question is usually, "What was it on the Richter scale?" The seismologist has two choices. One is to mentally redefine *Richter scale* to whatever is most convenient, and the other is to take the time to explain that the Richter scale doesn't actually exist—or, to be more precise, that *Richter scale* is a journalist's phrase, never used in scientific discourse. I once received a call from a reporter at the *Sunday Times* (of London) who asked for "a picture of the Richter scale." I replied, perhaps somewhat acerbically, that this was like asking for a picture of the Fahrenheit scale, and what did she want it for, anyway? It turned out that she wanted it for the paper's astrology page—to illustrate Libra. They were tired of having a picture of the usual balance scales and wanted some other sort of scale for a change. Charles Richter, who was not always the most tolerant of individuals, would not have been amused.[1]

Her misperception was that the term *Richter scale* referred to some piece of lab equipment, thinking perhaps of bathroom scales—something on which you could sit an earthquake to be measured, so that it would go, "Ping! 6.5!"

This raises a basic question—how do you measure the size of an earthquake? And what does size actually mean in the context of earthquakes, anyway? If we say that the magnitude of an earthquake is 6.5, what does that really mean—6.5 whats? What are the units? What are we measuring?

The idea that earthquakes come in different sizes goes back a long way. Early writers were quite capable of stating that one earthquake was a great one, another was less terrible, another quite a mild one. John Milne made an attempt to codify such language into three classes of earthquake: those producing sufficient force to produce cracks in buildings and throw down chimneys (class I); those sufficient to throw down, in addition to chimneys, some walls and weak structures (class II); and then rather a jump to class III—earthquakes causing general destruction.[2]

Milne's scheme was by no means the first.

DRAKE'S IDEA

In 1811 one of the most extraordinary earthquakes in history struck what is now the central United States. At the time the Mississippi valley was far from being "central"—it was on the frontier of expansion. As settlers moved westward across unfamiliar wild country, they naturally followed the main rivers, which provided the easiest method of transport. Towns sprang up along the banks of rivers like the Ohio and Mississippi.

The countryside around the Missouri boot heel—that bit of the state that sticks out southward into Arkansas, between the Mississippi and White Rivers—is flat and unprepossessing, almost featureless farmland. Nothing about this landscape would make you think of earthquakes, yet the clues are there if you know what to look for. Stand on one of the farm tracks that run across the fields, dead straight for miles. On either side of the track there will likely be a drainage ditch. Look along the side of the ditch and you may see a strange thing—the muddy slope is dark brown, then ahead of you it changes suddenly to a light sandy color, then dark again, then light again.

It looks even more remarkable from the air—the dark muddy color of the soil is marked by large circular patterns of the paler sandy color; the color changes you see in the ditches are where the tracks run in and out of these circular sand patches. These are in fact the traces of earthquakes. Very strong shaking can turn sand into a liquid—into quicksand, in fact. Where you have a layer of sand overlain by a layer of clay, during an earthquake the sand liquefies and then is forced up through the clay, where it erupts at the surface into great fountains of sand—and these have left their marks all over the Missouri boot heel.

The first of these earthquakes—at least, the first in the historical period (there are traces of prehistoric ones to be found as well) took place on December 16, 1811, at about 2:15 A.M. local time, followed by a second strong shock six hours later. A second great shock occurred on January 23, 1812, and yet another on February 7, 1812. The four quakes left a swath of damage from Memphis to Saint Louis; the little town of New Madrid was totally destroyed, and this town has given its name to the earthquake sequence.

To the north a Cincinnati physician named Daniel Drake kept a record of the earthquakes he felt. He had an interest in natural history, and when he published a book about Cincinnati and the surrounding countryside in 1815, he included his account of the earthquakes as an appendix.

He wrote of the first main shock,

The motion was a quick oscillation or rocking, by most persons believed to be west and east; by some south and north. Its continuance, taking the average of all the observations I could collect, was six or seven minutes. Several persons assert that it was preceded by a rumbling or rushing noise; but this is denied by others, who were awake at the commencement. It was so violent as to agitate the loose furniture of our rooms; open partition doors that were fastened with falling latches, and throw off the tops of a few chimnies in the vicinity of the town.[3]

Evidently Cincinnati got off lightly compared with the towns farther west.

Drake was not content to just list the dates and times of the earthquakes along with a few descriptive remarks. He wanted to categorize them. "The violence of different earthquakes, is best indicated by their efficiency in altering the structure of the more superficial parts of the earth, and in agitating, subverting or destroying the bodies which they support,"[4] he wrote. He proceeded to divide the different shocks into five classes. The first class consisted of the three main shocks, which "occupy above the rest, a decided elevation."[5] He then assigned three further events to the second class. The fourth class consisted of all the shocks that were felt only by people at rest; the fifth class were those only detected by pendulums or by a few sensitive people lying down. Everything between the fourth class and the six earthquakes making up the first and second classes made up the third class.

So for Drake the answer to the question of what to measure in an earthquake was the violence. There are no units for violence, so a classification is all that is possible. The problem, and Drake seems to have been aware of it, is that violence varies by place. He was well aware that the shocks that were merely strong in Cincinnati were far more severe farther west—and that they were significantly less strong away from the rivers altogether.

So while Drake's system looks superficially like Milne's, a classification of earthquakes that gives an idea of relative greatness, Drake was actually trying to show how strong an earthquake was, not how big it was. Drake's classes are a record of how strongly the shocks of the New Madrid sequence were felt in Cincinnati.

We now refer to this as *intensity*. The intensity of shaking is a shorthand classification for what Drake called the violence of an earthquake, or, as in Drake's fourth and fifth classes, the lack of violence. And as there are no units for violence, the best approach is to describe it in terms of the observable effects.

Such an approach is familiar in the measurement of wind speed—the Beaufort wind scale, devised by Rear Admiral Sir Francis Beaufort of the Royal Navy in 1805. Even without an instrument to record wind speed, one can still observe the effects of the wind on trees, leaves, flags, and so on, and assign a number between zero (no wind) and twelve (hurricane). The stronger the wind, the more pronounced the effects, up to and including damage to buildings, which begins around force 7 with a few loose tiles blowing off roofs, leading up to collapse at force 11 or 12.

Drake had the beginnings of an intensity scale, but he provided definitions for only the two lowest classes—class 4, not felt by people moving around, and class 5, felt only by highly sensitive people. All we can tell about the other classes is that 3 is stronger than 4, 2 is stronger than 3, and 1 is the strongest—even though one could infer something from the description he gives of the earthquakes. If the first shock tumbled only a few chimneys in Cincinnati, it would have been right off the scale in New Madrid itself, where the whole town was destroyed, so clearly Drake's class 1 is not the uppermost limit of earthquake violence.

The nearest anyone had come to this previously was in 1783, when one of the most destructive of all Italian earthquakes hit the Calabria region in the south of the country. Damage was heavy; many towns and villages were ruined, and in some places huge landslides changed the appearance of the countryside. After the earthquake the Academy of Naples sent a team out under the leadership of Michele Sarconi to survey the effects. This was the first time a scientific investigation of this sort had ever been organized in Italy. One result was the publication of a large collection of engravings the following year, including a map showing the distribution of damage, drawn up by the cartographer Padre Eliseo della Concezione. For this, Concezione created a simple classification of damage: towns undamaged, towns partly damaged, and towns destroyed. Each class had its own symbol (little drawings of towns in an appropriate state of devastation), and a glance at the map shows how far the damaged area extended.[6]

Concezione went further than Drake by mapping the spread of these primitive degrees of intensity, rather than just recording the effects at a single place, but Drake's scale included effects below the level of damage, even if his top three classes were virtually undefined. It would be a little longer before someone came up with something that combined a well-defined classification of earthquake effects with a map, and that breakthrough was made in a most unlikely place: Belgium.

THE EVOLUTION OF SCALES

Legend has it that a competition among British newspapermen to find the most boring headline ever was won by the entry "Small Earthquake in Chile: Not Many Dead." I was at a conference recently where one presentation ended with a slide that asked, "Would you like to know more about small earthquakes felt by Belgians?" This has to be one of the most rhetorical questions ever. Yet Belgium does have earthquakes, and they can be damaging, if not on the scale of major disasters. There is even geological evidence that in prehistoric times what is now Belgium suffered earthquakes stronger than any that have occurred in the historical record.

In 1828 a moderate shock occurred in the southern part of the country, causing some damage and attracting the attention of the German mathematician Peter Nikolaus Caspar Egen. Egen published a study in which he assigned numbers from one to six to describe different strengths of shaking, and he used these to plot a map that encapsulated in one figure the whole effect of the 1828 earthquake. The seismic intensity map was born.[7]

It often seems in science that if someone thinks of an idea before the world is ready for it, it will sink into obscurity. (Think of ancient Greek steam engines.) Such was the case with Egen's intensity scale.

It didn't catch on. It was another fifty years before the time was right for intensity scales, and then it was so right that two scientists reinvented the intensity scale independently—Michele de Rossi in Italy and François Forel in Switzerland. Once they realized they were thinking along the same lines, they collaborated and produced a combined version, known as the Rossi-Forel scale.[8]

Unlike the previous scales, and like the Beaufort wind scale, the Rossi-Forel scale was designed as a general-purpose tool, running all the way from the most feeble, barely perceptible shocks (intensity 1) to total destruction (intensity 10). The Rossi-Forel quickly caught on and was widely adopted by scientists all over the world. It was also adapted—quite a few people thought they could improve on the original, and as a result the story gets a bit complicated, not to mention confusing.[9]

One person who decided to improve on the work of de Rossi and Forel was Giuseppe Mercalli, an Italian volcanologist who published his own version of the Rossi-Forel scale in 1902.[10] It certainly was an improvement—he kept the number of degrees the same but lengthened the descriptions so that it was easier to tell exactly what, say, intensity 5 was actually like to experience. So, because the scale was more descriptive and easier to use, it became more popular than the original.

However, read articles about earthquakes on the Internet and you will repeatedly come across references to the "twelve-degree Mercalli scale." But Mercalli's scale had only ten degrees. So how did that come about?

The twelve-degree idea was the brainchild of the Italian physicist Adolf Cancani.[11] He noticed that both the Rossi-Forel scale and the Mercalli scale had an unfortunate gap at the top end. It jumped from "partial or total destruction of some buildings" at intensity 9 to total destruction of everything at intensity 10. So he suggested this should be spread out a bit: make total destruction degree 12, and then there was room for a better grading of levels of damage at intensities 9, 10, and 11.

That was Cancani's idea, but he didn't actually get around to writing the descriptions of damage at the different degrees. That fell to a German seismologist, August Sieberg. He rewrote the entire scale in 1912 and completely transformed it, lengthening all the descriptions immensely.[12]

This should have been called the Sieberg scale, which is what it was, but Sieberg called it something like "the Mercalli scale, modified by Cancani and formulated by Sieberg"—which was shortened to Mercalli-Cancani-Sieberg, or MCS, and in some countries, especially Italy, MCS is still in use today. (I have also seen it called Mercalli-Sieberg and even Forel-Cancani.)

Things got really confused when the scale was translated into English (in 1931) by two Californians, Harry Wood and Frank Neumann, who called it the Modified Mercalli scale. This misleading name stuck. Because this version was published in the prestigious *Bulletin of the Seismological Society of America*, it quickly became a standard, at least in the English-speaking world.[13]

However, as the saying goes, "God must like standards: there are so many of them." A whole series of spin-offs emerged as different writers attempted to improve the scale. And what tended to happen was that the authors all referred to these newer scales as Modified Mercalli. So if you read somewhere in an article about an earthquake that the intensity was 10 "Modified Mercalli," in fact, you won't really know what that means unless the author bothers to note "1931 version" or "1956 version" or "1980 version" or whatever, which seldom happens. And really careless authors leave out "Modified" as well, which is why you see erroneous references to the "twelve-degree Mercalli scale."

But back to Richter.

RICHTER AND THE PRESS

Charles Richter (1900–85) undoubtedly has the right to be regarded as the most famous seismologist in history. In the late 1920s and early

1930s he was working as a seismologist at the Seismological Laboratory of the Carnegie Institute in Southern California. He had grown up in Los Angeles with his maternal grandfather after his parents had divorced while he was still a small child, and he had attended Stanford University.[14] Now he was collaborating with Beno Gutenberg, a refugee from Germany who was working at the California Institute of Technology in Pasadena. Together they issued a regular bulletin of local earthquakes in Southern California. Recording the date, time, and place was fine, but they also wanted to distinguish between small, medium, and large events.

Richter saw that using intensity for this was a problem: it didn't take into account earthquakes that occurred far from human habitation. Suppose the earthquake occurred out in the desert? Or out at sea? For an offshore earthquake the strongest shaking would be that at the nearest point on land, and if the epicenter was far enough out to sea, the shaking on land might be weak.

Intensity didn't work for earthquakes out in the desert, since no one was there to experience them. But there was yet another drawback. Whenever reports came in from people who had felt an earthquake, deciding whether the reports matched one degree of the scale or another was largely subjective. Worse, sometimes the reports were contradictory. Two observers in the same town could easily have different experiences, and earthquake shaking is so capricious that observers in the same building often have different experiences.

Gutenberg and Richter wanted something that was objective and reproducible, and the physical records written by seismometers looked like they might provide the answer. They knew that larger earthquakes produce bigger seismic waves and that a large wave would get smaller and smaller as it spread out away from the fault, just as water waves become weaker and weaker with distance from a splash.

So together Gutenberg and Richter devised a system based on the maximum wave peak and adjusted for distance. They chose a "standard earthquake," defined as an earthquake one hundred kilometers from a seismometer that produces a maximum peak of 0.001 millimeters, and called this magnitude zero. Then, if the recorded peak was bigger, or the earthquake was farther away, the magnitude would increase according to a logarithmic formula. Using a logarithmic formula kept the numbers within reasonable bounds: if the signal was ten times bigger, the magnitude was greater by one, not greater by ten.

Richter chose the word *magnitude* after the stellar magnitude astronomers used to describe the size of stars. And the scale itself was named the *local magnitude scale,* because it measured the magnitude of local earthquakes in Southern California.[15]

Although Richter could be a crusty character (it has been suggested he may have been mildly autistic), he was more outgoing than Gutenberg, for whom English was a second language. Consequently, when journalists called after an earthquake that was large enough to be of interest to the public, it was always Richter who spoke to them. He would simply say something like, "This earthquake registered magnitude 4.5"—but then journalists would write this as "The earthquake was 4.5 on Professor Richter's scale." And this phrasing gradually morphed in the press reports into "4.5 on the Richter scale."[16]

That usage has persisted, so far as journalists and the public are concerned. But not seismologists! Since the technical name of the scale was local magnitude (ML), that's how seismologists refer to it in written work or when speaking to each other. Only when a journalist or member of the public phones up and asks, "How big was it on the Richter scale?" will the seismologist either say, "Well, actually, it's not really called the Richter scale" or, much more commonly, swallow any scruples and say "4.5" because that's so much easier than going into a long explanation.

THE SCALING OF SCALES

One major difference between an earthquake magnitude scale and other scientific measurements is that it has no units—we don't talk about "4.5

Richters," for instance. The reason for this is straightforward: there's no actual property of the earthquake that the local magnitude scale measures. In calculating the magnitude, a seismologist measures the maximum wave peak—that is, the width of the biggest wiggle on the seismogram—calculates the distance of the earthquake from the seismometer, and plugs these two different values into Richter's formula. The number that comes out is not a distance or a weight or a force. It's just a number. But it's a useful number, because it tells you how one earthquake compares with another and provides the basic ranking of "large, medium, and small" that Richter was after.

Another factor is the huge variation of scale that earthquakes display. Although Richter's scale was logarithmic in steps of ten related to the displacement on a seismogram (which is what the seismologist is actually measuring), it subsequently emerged that the scaling in relation to seismic energy released (which is the real destructive power of the earthquake) was more like steps of thirty. So imagine what would happen if we called an earthquake of magnitude 1 "1 R" for "1 Richter." A magnitude 2 earthquake would be 30 R; magnitude 3 would be 30 x 30 = 900 R, and when we get into the damaging range at magnitude 6, we are up to 24.3 million R, which is a cumbersome number. You can see, incidentally, why it's wrong to imagine that a magnitude 6 earthquake is only 20 percent worse than a magnitude 5; in energy terms, and therefore potential to do damage, it's thirty times worse.

Unfortunately for Richter and Gutenberg, they discovered that when you calculate the magnitude of a particular earthquake from different stations, you get different answers. This can be for a variety of reasons—it may be that when the fault moved, more seismic energy went in one direction than another. Or perhaps the local geology around one recording station has a tendency to amplify shaking. Or it could just be caused by the general complexity of all the seismic waves emitted by the earthquake.

The solution to this problem was simply to add up all the magnitudes from each recording station and divide by the number of stations.

As Richter wrote, "Lest the impression should be created that great precision is being claimed for this method, it is desirable here to emphasize its actual crudity."[17] This is why you generally see magnitudes quoted to only one decimal place: 4.5 and not 4.52. Finer resolution is really meaningless. In fact, Richter often tended to quote magnitudes only to the nearest quarter-degree.

This system of calculating earthquake size also explains another puzzling phenomenon: Why does the magnitude keep changing in news reports about an earthquake? The first news bulletin gives one number, then the next one cites a different figure, then yet another one the next day. This is simply because the number representing the magnitude is an average of all the stations' magnitudes that have been reported, and as more are collected the average changes.

So far so good. But while the local magnitude worked nicely for local earthquakes in Southern California, it soon became clear that it didn't work for large, distant earthquakes. You could record an obviously huge earthquake in Chile or Alaska or wherever and the magnitude calculated from Richter and Gutenberg's formula would come out the same—somewhere around 6 to 6.5. This problem is known as saturation, and it is related to the maximum response of a seismometer. Just as a sponge gets to a point where it can't absorb any more water, no matter how much you pour onto it, the magnitude scale simply stops going any higher, no matter how much larger the earthquake is.

Gutenberg and Richter did not want to change the existing scale, which was working nicely for the job for which it was designed. Instead they started looking for a different way to calculate magnitude that would extend the scale upward, while still producing numbers that were compatible with local magnitude. Gutenberg, who increasingly took the lead, decided one could get a more controlled and consistent result by always measuring one specific type of wave. Two avenues looked promising. One was to measure the amplitude of only the first few waves at the beginning of the seismogram and make the calculations from that. This became known as body-wave magnitude (abbreviated mb). The other was to concentrate on the surface waves at the end of the seismogram and take the measurement there, giving surface-wave magnitude (Ms).

Gutenberg focused particularly on the second approach, tweaking the formula so that the numbers produced by the surface-wave scale looked as though they had been tacked onto the top end of the local magnitude scale—in other words, the two scales fitted together at around magnitude 6 without an awkward gap. Gutenberg designed the surface-wave magnitude scale to be an extension of the local magnitude scale; now it was possible to provide magnitude values for large earthquakes happening anywhere in the world. Magnitude was no longer limited to moderate-sized quakes in California.

JUST A MOMENT

There were still more problems. It soon became clear that using surface waves to measure deep earthquakes would be difficult. If the focus of an earthquake was deep, surface waves didn't develop strongly, so these large, deep quakes got shortchanged. Surface-wave magnitudes weren't immune to the saturation problem either; it just set in higher up the scale. Once you got to a bit over magnitude 8, all earthquakes seemed to be the same size, even when their effects varied greatly.

A solution was not found until much later. In the late 1970s, a Japanese-American seismologist working in California, Hinoo Kanamori, came up with a solution to the problem that magnitude was only a measure of a wave produced by the earthquake, not any real property of the earthquake itself. There was one property of an earthquake that actually could be measured, Kanamori realized, and it could be used as a measurement. This was the work done, or, more specifically, the "seismic moment."

In physics *work* means something a little different from its everyday definition. Put simply, it is mass times the distance moved. Think of trying to shift a heavy refrigerator from one room to the next, and compare that effort to carrying a packed suitcase for several blocks. Probably both involve about the same amount of human effort in total—this is the work done. Moment in physics is a similar concept; it considers force applied over distance in a specific direction (carrying the suitcase north rather than east).

So to go back to what an earthquake actually is—the abrupt movement of rocks along a fault—it's possible to measure the amount of displacement on the fault and the area of the fault plane over which the displacement occurs—and this gives you the seismic moment in Newton-meters (the scale used for measuring moment), which, at last, is a property of the earthquake and nothing else, and is a true measure of an earthquake's size.

The downside is that once you calculate this, the numbers are rather large, and it's a bit difficult to communicate the size of an earthquake if you have to say, "This earthquake measured 17,500,000,000,000,000,000 Newton-meters." Kanamori decided to convert these huge unwieldy numbers into values that approximated those in existing magnitude scales.

This turned out to work rather well, and the result was named the moment magnitude scale, with the abbreviation Mw, occasionally referred to as the Kanamori scale.[18] The great advantage of Kanamori's invention was that it didn't saturate at all and worked perfectly well from the smallest to the biggest earthquakes. So instead of having two scales, one that worked fine for small earthquakes up to about 6, and another that started at about 5 and went up to 8, one scale now covered the whole range, even up to the biggest earthquakes with magnitudes greater than 9.

However, the values of the new scale did not completely agree with the older ones. Moment magnitude and surface-wave magnitude agree pretty closely in the important range between 5 and 8, but moment magnitudes are consistently smaller than the equivalent local magnitudes for the same earthquakes. So something that was 6.0 ML might be only 5.7 Mw.

This remains a perpetual source of confusion. Though most seismologists agree that Kanamori's scale is the best one to use, it is harder to calculate, and, particularly in countries that don't have large earthquakes, it's easier to keep using local magnitudes. This is not a problem for seismologists, who see "6.0 ML" and immediately understand. But for anyone who imagines that there is a unified, shiny, inviolate system called the Richter scale, it all becomes a bit confusing.

So next time you hear an earthquake reported on the news, listen carefully to exactly how the magnitude is described.

BACK TO INTENSITY

For a while it seemed as though magnitude had trumped intensity as a tool for measuring earthquakes. However, this was to neglect that magnitude and intensity measure two different things—how big, how strong—and that intensity is still useful as a measure of how strong an earthquake is. Because it directly expresses the damage caused by the shaking, anyone can tell from seeing a value of intensity 7 that this means moderate damage to buildings.

Intensity information is immediately useful, for instance, to insurers, whose prime interest is in the amount of damage for which they might have to foot the bill. Also, intensity information is easy to gather over a wide area. The old method was to distribute paper questionnaires throughout the area affected by an earthquake. People could answer questions like "Did your windows rattle?" "Was there any damage?" and from this a seismologist could assign intensity values and make a map showing the whole distribution of effects, with the highest intensities in the middle, usually corresponding more or less to the epicenter, and gradually becoming weaker with increasing distance. To make the map easier to read, seismologists as long ago as the late nineteenth century would draw contour lines (known as isoseismals) to mark out where the intensity was greater than 4, greater than 5, and so on. The most modern intensity scale is the European Macroseismic Scale (*macroseismic* is the technical term for observations of the effects of earthquakes). This was published in 1998 after almost ten years of development and, despite its name, has been widely used around the world.[19]

Nowadays the main means of gathering such information is through the Internet. Most developed countries have a national web page where anyone can fill in a questionnaire online, and the map of all the data is continually updated as more and more replies stream in. This is an incredibly helpful way of visualizing how an earthquake is experienced over a wide area, and such sites are extremely popular after

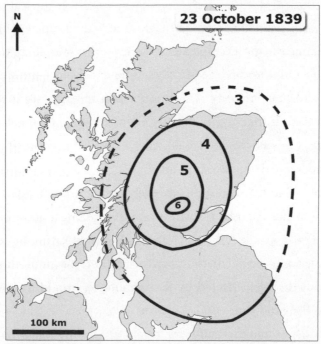

An example of an isoseismal map for a small- to moderate-sized earthquake; this is actually the 1839 Comrie earthquake that started the British Association's earthquake investigations. The numbers indicate the severity of shaking within each contour: 6 = slightly damaging, 5 = strong (objects fall over), 4 = generally noticeable (windows rattle), 3 = weak (few people feel it).

a strong earthquake—especially the "Did You Feel It?" site run by the USGS, which produces maps of most major earthquakes anywhere in the world (http://earthquake.usgs.gov/earthquakes/dyfi/).

A surge of interest in historical earthquakes since the late 1970s, largely to study seismic hazard (see Chapter 11), also led to a renewed interest in intensity. Intensity has the advantage of allowing you to gather data retrospectively, which you obviously can't do with instrumental data. If I want to study an earthquake that happened in the seventeenth century, it's possible for me to dig up contemporary descriptions from libraries and archives, interpret them as intensities, make a map, and compare it to maps of modern earthquakes. And in this way historical seismologists started estimating magnitude for past events.

It's possible, for example, to compare maps of the effects of the 2010 Haiti earthquake with a reconstruction from historical accounts of the effects of the previous earthquake there in 1770. The distributions are similar, but in the earlier earthquake strong shaking was more widely distributed. So while the 2010 earthquake was felt only slightly in Jamaica, the 1770 earthquake was quite noticeable. So if the 2010 earthquake had a magnitude of 7.0, one can estimate at once that the 1770 earthquake must have been larger—perhaps 7.5. With enough data this can be turned into a formula for calculating magnitude from an intensity map, and the results are actually quite reliable—in fact, not much worse than those you would get from seismograms, providing you have a reasonably good historical picture. Seismologists often are asked how anyone can know the magnitude of an earthquake in historical times when seismometers have been around for only a little more than one hundred years. This is how.

Actually, we have two approaches to estimating magnitude from intensity data for historical earthquakes. One is to look at the maximum intensity, the other, the overall area affected. The first of these is not reliable for a number of reasons. I have already mentioned the problem of an earthquake that occurs in a desert without any observers. For smaller earthquakes the depth at which the fault breaks affects the intensity—if the focus is shallow, buildings directly above it get the full kick, and the maximum intensity is higher than it would have been otherwise—so small shallow earthquakes look larger than they really are. Another factor is building quality—in an area where the houses are of very poor quality, they all come crashing down without particularly strong shaking, so one can't really distinguish different degrees of damage. In contrast, the size of the area affected is a robust way of estimating the size of an earthquake, so this is a much better thing to use.

FLYING ROCKS

Intensity is a measure of shaking, but it's only a descriptive rating. Physical measurements can also be made, and these are handy for engineers in calculating how building designs will fare in an earthquake, something I will discuss in more detail in Chapter 10.

Although seismometers record the physical movement of the ground, shaking so strong as to cause damage will send the instrument way off scale. Therefore seismologists use another type of instrument, called an accelerometer, which is essentially a ruggedized seismometer designed to make accurate records of very strong ground movement. There are three main ways in which that movement can be expressed.

How much did the ground actually move—say, in centimeters? This measures the ground displacement, and the maximum value is referred to as the peak ground displacement.

How fast did the ground move? This can be measured in centimeters per second and is referred to as ground velocity; the maximum value is the peak ground velocity.

How quickly did the ground go from not moving at all to its fastest speed? This is acceleration (in centimeters per second per second), and the maximum is the peak ground acceleration.

All of these are important. The ground could move with a huge displacement, but if it happened very gradually, it wouldn't be very damaging. Or it could move very fast, but if the displacement was tiny, that wouldn't be so bad either. The one that engineers use most often is acceleration, since that can be directly related to the force acting on the building, which design calculations need to take into account. So if you see references to "PGA" in discussions of earthquake hazard, it has nothing to do with professional golfers, just peak ground acceleration.

While it's possible to measure PGA in centimeters per second per second, it's more common to measure it as a proportion of the acceleration due to gravity, written as g. It used to be drummed into school physics classes that the acceleration of gravity is thirty-two feet per second per second; now it's usually written as 981 centimeters per second per second (or cm/sec/sec, or cm/sec^2). Writers describing the flights of test pilots or astronauts often refer to the pilot experiencing so many gees; it's exactly the same thing. A jet aircraft undertaking fancy maneuvers undergoes strong acceleration, which can be unpleasant for the pilot. Whether it's a plane or the ground during an earthquake, gravity is a unit for describing acceleration.

Some seismologists used to think that the physical limit to the maximum PGA an earthquake could produce was a value of about 0.5 g. Other seismologists disagreed, pointing to reports in some large earthquakes, notably the great 1897 Shillong earthquake in northeastern India, of stones being flung into the air. If an object is thrown in the air, it must be receiving an upward acceleration greater than the downward force of gravity; in other words, the vertical acceleration must be greater than 1 g. The first group of seismologists then defended their position by casting doubt on the observers in Shillong. Maybe they were mistaken. Maybe they only thought they saw stones flung in the air. Sadly, this is a common trait in scientists; if an observation made by an ordinary individual clashes with a cherished theory, blame the observer for being "untrained." In the eighteenth century the same sort of argument would be trotted out to show that meteorites didn't exist. There are no stones in the sky; therefore stones cannot fall from the sky; therefore anyone who says they saw a stone fall from the sky is an "untrained observer" and we can forget about him.

Then in 1971 an earthquake occurred in the San Fernando area of Southern California and a PGA of 1.17 g was recorded. That settled that argument. Since then ground accelerations even greater than 2 g have been recorded. Probably there is a physical limit, but no one knows what it is.

BIGGEST AND STRONGEST

Magnitude and intensity represent two quite different measurements—overall size (of the whole earthquake) and severity of shaking (at a particular place). Unfortunately the public tends to use these two measurements interchangeably. One common problem in California occurs when people feel a moderately large earthquake at some distance. A few ornaments rattle, they hear that the earthquake measured 6.5, and they say, "Huh! We've lived through a 6.5, and it was no great shakes!" They then start discounting the danger of earthquakes. But then a 4.0 event happens nearby and feels much stronger—"Hey! That was supposed to be only a 4.0, and it felt much worse than a 6.5! What gives?" Answer: they are confusing magnitude and intensity and do not realize that the strength of shaking diminishes with distance. Size is not everything.

"The open-ended Richter scale" is one phrase that adds to the confusion. Whenever I hear this, I ask if the speaker also refers to the "open-ended Fahrenheit or Celsius scale" for temperature. Perhaps people describe magnitude as open-ended because intensity is not— like the Beaufort wind scale, the number of defined degrees or classes is finite, and you can't go outside these—nor can you meaningfully have fractions.

Then there is the sort of inquiry I get on the phone at least a few times per year. An engineer calls and asks, "We've been told to design this building for earthquake force 7. What does this mean?" To which I reply that it doesn't mean anything, and the engineer should go back to the client and ask them to explain themselves better. However, it probably means intensity 7, since a magnitude 7 earthquake would mean something different depending on how far away it was. Magnitude and intensity often get confused—yet the difference is not hard to understand.

Occasionally prophets of doom try to scare people with predictions of terrible earthquakes to come and promise earthquakes with magnitudes much greater than any ever recorded—magnitude 12, even. This is just thoughtless—where is all that energy going to come from? Current thinking among seismologists is that earthquake magnitude has a physical limit, and that is about 9.5 Mw, or about the size of the 1960 earthquake in Chile. No fault could accumulate enough strain to produce a magnitude 10 earthquake—it would break before it ever got to that point.

So if the size of earthquakes has an upper limit, what would be the lower limit? Richter started with a "standard earthquake," which he called magnitude 0, but this was not the smallest event that could ever be recorded. With smaller earthquakes we go into negative numbers, so it is possible to record an earthquake with thirty times less energy than a magnitude 0, and this will be magnitude −1. Such an earthquake is too small for anyone to feel, but modern instruments close enough to the fault can pick it up.

The smallest earthquake recordable with modern instruments (at close range) is about −3. An easy way to visualize a −3 quake is to imagine you have half a brick in your hand. Stretch your arm out straight at shoulder height and drop the half-brick. The impact it makes on the ground is about the same energy as a magnitude −3 earthquake. So from −3 to 9—twelve degrees of magnitude—one goes from a trivial bump to the devastation of a massive earthquake that rips the ground for a thousand miles and changes the landscape.

Reminiscing about the magnitude scale thirty years later, Richter said that given how crude the scale was, it was remarkable that it worked at all. And the only reason it worked was the immense range in power from the largest to the smallest earthquakes. Crude the scale might be, but the difference in energy from the small everyday bumps and bangs to city-destroying major catastrophes is so large that Richter's invention has proved a lasting success.

6

THE WAVE THAT SHOOK
THE WORLD

THE IMPACT OF EARTHQUAKES ON OUR PLANET IS PRO-
found. The same tectonic processes that produce earthquakes are also
responsible for throwing up great mountain chains like the Rockies and
Alps. The landscape we see around us is a slow battle between the forces
of erosion, wearing away hills and mountains, and tectonic forces pushing them up again. The effects on human society are also widespread.
Every death is a personal loss, even when a minor earthquake kills only
one or two people. Larger and deadlier earthquakes have a progressively more serious effect on whole communities, and in numerous
cases people have abandoned entire towns or villages after a major
earthquake and rebuilt somewhere else. One example is Noto in Sicily.
Destroyed in 1693 by the same earthquake that engaged Flamsteed's
attention (see Chapter 2), after the earthquake it was built afresh on
another site. The overgrown ruins of old Noto can still be seen today.
Walking through them at dusk is a spooky experience.

There has even been speculation that some ancient societies were
so weakened by the effect of a disastrous earthquake that they collapsed
altogether, though it's tough to prove from archaeological evidence
alone.[1] It would not be surprising, small city-states, such as those that

flourished around the Mediterranean during the Bronze Age, had limited resources with which to cope with a natural calamity striking their small territory. Weakened by an earthquake, a small city-state would be easy prey for any aggressive neighbor.

If one were going to pick the three most significant earthquakes of all time, the ones that changed the way we think about earthquakes, most seismologists would not have much difficulty in agreeing on a list. Two of them we have already met. The great Lisbon earthquake of 1755 killed off the idea that earthquakes are sent by God (though, like many superseded ideas, it is taking a long time to finally expire) and also led to a huge burst of scientific interest in earthquakes. The San Francisco earthquake of 1906 has become the iconic idea of an earthquake in the popular imagination, and it caused a seismological revolution through Reid's study of elastic rebound. One from Europe in the eighteenth century; one from America in the twentieth century. Now it is time to bring in the third of the trio: Asia in the twenty-first century.

When a seismologist is woken up by the telephone ringing early in the morning, it usually means bad news, and if it is a public holiday as well, it is definitely very bad news. So when I was woken early on the morning after Christmas Day 2004 (in the United Kingdom this is the Boxing Day holiday), I immediately suspected that a disaster had occurred somewhere. What was curious was that this was a repeat of the previous December 26, when a strong earthquake had flattened the ancient city of Bam in southeastern Iran, with terrible casualties (see Chapter 1). This time, though, things would be much worse.

A colleague of mine was on the line. "There's been a big earthquake in Sumatra," he said. "USGS is saying low 8s, could be 8.5. Could you come in?"

Soon I was dressed and driving toward the office through the dark and silent streets of Edinburgh, mentally reviewing what I could remember about the tectonics of Sumatra. Once I arrived and got the computer up and running, one of the first things I did was check on the aftershocks. After a major earthquake aftershock activity tends to be intense, and these smaller settling-down shocks trace out the area of the fault that has just moved and where the crust is still not entirely stable in its new, postquake configuration. I downloaded the epicenters and threw them into a visualization program of mine. What I expected to see was a cluster of red dots somewhere on the west coast of Sumatra. What I saw was something else entirely, something that was at first puzzling and then shocking. The line of red dots started on the Sumatra coast—but then went all the way up to the Andaman Islands, far to the north.

Aftershocks that distance north? That meant that the fault must have ruptured for more than a thousand kilometers, something I had never seen in my working life. Earthquakes of magnitude 8 don't do that. Only one explanation was possible: this was magnitude 9.

Back in the days when seismologists used mostly surface-wave magnitude to measure the size of earthquakes, it was impossible to measure an earthquake larger than about 8.5, as the scale simply stopped functioning at such heights (the saturation problem discussed in Chapter 5). With the introduction of moment magnitude in the 1970s, this was no longer a problem, but no earthquake of magnitude 9 had occurred since then. Retrospectively, only two earthquakes had ever been reclassified as 9s, the quakes in Chile in 1960 and Alaska in 1964. So most seismologists in the age of modern computing and the Internet had never handled such large earthquakes and tended to think of the high 8s as the practical top end of the magnitude range.

The Sumatra earthquake of December 26, 2004, proved to be a rude awakening and a severe test of the tools and techniques seismologists were using. The fault started to break near the northern point of Sumatra and then tore open in a northward direction. The vast length of the rupture meant that the initial shock waves from near the epicenter had been traveling for several minutes before the waves from the far end of the fault had even started. This produced a horrendously complex series of overlapping and interfering wave patterns, which proved extremely difficult to analyze correctly. This was one of the reasons why the initial magnitude estimates were all too low.

But if the great Sumatra earthquake of 2004 forced seismologists to think again about how to handle extremely large earthquakes, it changed the way the public thought about earthquakes entirely, and even changed the language people used to describe them.

And that was because of the tsunami.

Before 2004 not many people could have said what a tsunami is. Media reports of earthquakes sometimes mentioned a "tidal wave," a term that always irritated seismologists because they have nothing whatever to do with tides. After 2004 "tidal wave" has almost vanished from use. People know what a tsunami is now.

Christmas is holiday time, and for those fed up with spending the season in the grip of a European or North American winter, the beaches of Thailand make an attractive alternative. On that morning the most popular resort beaches were packed with tourists enjoying the sun. Who among them would have had any idea that they were at risk of a natural disaster? Where could the danger come from on a fine day with no risk of typhoon? For many, the first sign that something might be amiss was the approach of a wave from out at sea that seemed rather larger than usual. Then it became clear that not only was it large, it was fast, and it kept on coming.

The wave surged up the beaches, rolling over anyone and anything in its path, and kept on moving, up over the back of the beaches and into the streets of the resorts themselves, surging over walls and across boulevards with a frightening roar, surging into buildings, picking up street stalls, tossing cars and trucks aside as if they were toys. Everywhere screaming people were running that way and this, looking for escape. Some found refuge on the upper floors of buildings, others tried climbing trees to get above the swirling debris-choked waters. Thousands never made it.

None of those fleeing tourists had any idea that an earthquake had just occurred. Thailand was too far away from Sumatra for people to have felt it. But all along the fault line, a wave of water had been set in motion that would cause destruction when it reached the nearest shore.

Back in Edinburgh, as we watched the news reports with horror, the colleague who had first phoned me turned to me. "This time last year, I was on that beach," he said.

The northernmost coast of Sumatra, around the city of Bandeh Aceh, was swamped almost immediately after the earthquake. North of Sumatra, the fault runs almost north-south, and the wave launched eastward, straight at the tourist resorts of Thailand. But the wave was traveling westward as well, toward more tourist beaches on the coasts of Sri Lanka and southern India. The wave traveled straight across the Indian Ocean at high speed, arriving about two hours after the earthquake. The scenes of horror that had already unfolded in Thailand were repeated on the other side of the Bay of Bengal.

Except on one beach. There, as elsewhere, people noticed that the sea was starting to go out. It was like a low tide but more rapid, and the water went out a long way. This was puzzling to everyone except an English schoolgirl named Tilly Smith, who was enjoying a tropical Christmas with her parents. She remembered from her geography classes something called a tsunami, a destructive wave that sometimes accompanies earthquakes, and that sometimes before a tsunami strikes the sea recedes for no obvious reason.

Well, the sea was receding when it plainly had no cause to. When you learn about exotic things with Japanese names in geography class, you don't ever expect to encounter them yourself, especially when your school is in England. But what other explanation could there be? So she told her parents that really, this was probably not just some inexplicable thing that was happening, but that they were actually in a lot of danger. Fortunately, her parents believed her and started yelling at people to clear the beach.

Sure enough, a little while later the tsunami rolled in—over a deserted beach.

PHOTOS FROM THE DOOMED

Tsunamis became so much a part of public consciousness after the 2004 disaster that for a while, whenever any earthquake occurred, the first question journalists asked was, "Will there be a tsunami?" When an earthquake occurred in central Russia, as far from the sea as you could get, this question was easy to answer.

There were two other ways in which the 2004 Sumatran earthquake changed things: it used to be a general observation that Western tourists were practically never the victims of an earthquake disaster. Earthquakes with high death tolls usually involved the collapse of a lot of substandard housing in relatively deprived areas, and if any tourists were in the area (which usually they weren't), they would be in modern hotels that were reasonably well built. Sumatra changed all that—the beaches that the tsunami rolled over were crowded with Western tourists, and many of those killed left behind relatives who never imagined they might lose a family member to an earthquake.

The other is the vast amount of film and photography collected, preserving vivid images of what it was like when the wave crashed over the beach. I always used to say, when giving public lectures, that we have few good pictures of tsunamis because, by and large, if you see one coming, taking photographs is the last thing on your mind—you want to devote all your attention to getting to safety as quickly as possible. But the increased use of fixed cameras, such as security cameras, means that much gets captured on film today that would not have been a few decades ago. And in an area full of tourists, many with video cameras, it was inevitable that some people in safe spots, such as high-up hotel balconies, were able to record the whole scene.

But what are we to make of the Canadian couple who, as the tsunami surged toward them, stood on the beach and made no effort to run? Instead they just took photo after photo of the wave coming closer and closer. In the last frightening image the wave is almost on top of them. That photo shows what a tsunami looks like when it is a few feet away—something that no one has seen before and lived to recount. It is a picture of certain death approaching. The camera was found later amongst the debris left behind by the water, and the memory card could still be read. We know who it belonged to, as the couple appeared in some of the earlier photos on the memory card—typical vacation snaps.

What were they thinking as they stood there? Did they imagine the wave would sweep harmlessly around them? Or did they decide they had no chance of escaping? Whatever the explanation, at the cost of their lives they left some awe-inspiring pictures that may never be duplicated.

MOVING HOUSES

So how do tsunamis occur?

Physically a tsunami is nothing more than a big splash. Imagine throwing a rock into a lake—the rock enters the lake and displaces a volume of water equivalent to the volume of the rock. Because you threw the rock in the lake, the displacement is more violent than if, say, you just lowered it in gently. The water fountains up, and a wave ripples out in all directions.

A tsunami is essentially the same thing on a bigger scale, and an earthquake is not the only phenomenon that can produce one. A big meteorite falling into the sea is similar to you throwing a rock in a lake. It's the same process, except the scale is greatly increased and the impact would take place at a huge velocity, producing (if the meteorite were a big one) a massive splash and a huge wave spreading out in all directions.

Large meteorite impacts are rare events. Much more common, and still essentially similar, are landslides. This is something often seen in Norway or Alaska, where high rock cliffs loom over deep fjords. Landslides and rockfalls are quite common there, and when a shower of rocks comes down, it can make an impressive splash. The wave in the confined space of the fjord can reach spectacular heights, washing up the valley sides and scouring away the forest.

The shaking from an earthquake can also dislodge rocks from a cliff. This happens quite frequently in mountainous districts where rocky slopes are weak from long exposure to rain and snow, and it doesn't take much to destabilize them. When earthquakes hit cities, clouds of dust arise from collapsing buildings, but when earthquakes happen in hill country, clouds of dust arise from rocks as they shower down from the steeper slopes. If an earthquake strikes a mountainous coastal district, the result can be a landslide that falls right into the sea, causing a huge splash—and a tsunami. This has been observed on several occasions in Alaska, which has the requisite combination: cliffs, sea, and earthquakes.

But something similar can happen in places without mountain cliffs towering over the sea. Landslides don't have to occur on hill slopes; they can occur on underwater slopes as well. The effect is the same; even though the movement of rocks is entirely underwater, it still displaces a large volume of water, which creates a wave.

The tourists who packed the beaches of Thailand on that December morning in 2004 never imagined they might be in any danger from a tsunami, for who had ever heard of one in Thailand before? But this wasn't the first time a tsunami surprised the coast of an unlikely country.

Burin Bay is a sleepy little place in the southeast of Newfoundland. The coast is a rocky one, full of twisting little inlets and small islands. The settlement faces the sea and lives off the sea; it has been a fishing port since it was first settled in the early eighteenth century. Access by land today is along a winding minor road, but until the 1960s there was no road at all, and the only access was by boat. The timber houses are scattered throughout the landscape like a handful of dice, and the "streets" are tracks connecting one bunch of houses with the next, from one cove to another.

In 1929 the Emberley family was living in Burin. On the winter's evening of November 18, daughter Louise, a young woman of twenty-three, was watching her mother bake apple dumplings in the kitchen when she heard a strange noise. At first she thought something was wrong with the stove. But when she and her mother ran outside, they noticed that the ground was trembling beneath their feet. Strange, but it passed quickly with no harm done. Back they went into the house.

After the evening meal Louise went for a walk to the post office to see what news there was. In the absence of radio or television, the villagers' main source of information about the outside world was the telegraph. The telegraph operator would receive news reports and either write them out and stick them on the wall of the post office, or simply read them out if anyone was around to listen. In this way the inhabitants of Burin would learn what was going on outside the village.

Louise had a choice of two routes to take; instead of walking along the upper road, she took the path that ran down by the shore, much of it a walkway over rocks and pools with planks for bridges. As she walked, she was astonished to see the harbor turn dry. The water just disappeared, leaving the boats moored there, stuck in the mud, leaning on their sides at crazy angles.

By the time she got to the post office, a large group of people had gathered to find out what news there was of the earthquake they had felt. Louise had been there for a only few minutes when everyone heard a loud rushing noise coming from the shore. They turned to look—and what they saw horrified them. A wall of water was sweeping inland. It was a bright, still, moonlit evening, and the whole scene was clear to the eye: a wall of white, foaming water, higher than any wave they had ever seen. The noise grew to deafening levels as the water swept over the harbor, tearing buildings apart, pitching boats before it, and smashing anything in its way.

The group gathered around the post office had no idea what was happening; it seemed as if the land were sinking and the sea rolling in to fill the space. The one thing they knew was that they wanted to get to high ground. They ran.

Now all the villagers were gathered on the highest part of the slope. Louise's mother had come up by the high road and was safe, much to Louise's relief. As they stood there watching, the wave surged out again, only to collide with a second wave that was sweeping in—and then a third. Amongst the group was the local lighthouse keeper, the one man who understood what had happened. He explained that the earthquake they had felt must have been out at sea, and the wave was the result.[2]

He was right. Although Newfoundland is not exactly in an earth-quake zone, large earthquakes occasionally occur in the most unex-pected places. In 1929 it happened in the Grand Banks, a famous fishing ground off the Newfoundland coast, where the shallow coastal waters of eastern North America start to plunge down to the deep ocean floor. When a 7.1 magnitude earthquake struck on that November evening, it loosened the sediments on the sloping sea floor and caused a massive slump.

How massive? Think of where you live, and imagine a place one mile to the north of it. Now think of a place one mile to the east. Now think of a square with those two places at two opposite corners, and your home on a third corner. That's a square mile. Now imagine that area flipped up so it sticks one mile up into the sky; turn it into a cube, a mile on each edge. You could fit a lot of rock into that space, a cubic mile. Now multiply by fifty. That was roughly how much mud and silt rolled down the slope to the ocean bottom at a speed of about forty-five miles per hour. No wonder it made a splash.

The tsunami mixed up land and sea in bizarre ways. Boats were carried on shore and dumped high and dry, while houses were swept out to sea intact. Captain Abram Kean of the SS *Portia*, who arrived a few days later, described the scene: "Imagine our wonder and surprise on turning the point of the channel to be met by a large store drifting slowly along the shore seaward; then a short distance another store or a dwelling house until nine buildings were counted, strewn along the shores before the harbour was reached."[3]

A famous photo taken after the tsunami shows a large schooner in one of the bays near Burin, with a wooden two-story house in the water directly astern of it. The house belonged to a resident of the village of Port au Bras and was swept intact out to sea by the tsunami. The owner and his father set out to tow the house back. Eventually the house was set back in its proper place in the village.

The death toll from the 1929 tsunami was not high: twenty-nine, including one who died from injuries the following year. The area affected was not large and was sparsely populated. The tsunami was detectable on the other side of the Atlantic but only as the merest trace on tide gauges. It was a very different story from the 2004 tsunami, which struck the other side of the Indian Ocean with great force, even killing a few people on the east coast of Africa. Similarly the tsunami from the Lisbon earthquake in 1755 killed people on the coast of Brazil.

FLIPPING THE SEABED

So what makes some tsunamis deadly at long range, while others are purely local disasters? To understand this it's necessary to go back to a concept that I mentioned briefly at the end of Chapter 3—subduction.

Subduction zones are special. I once attended a seminar given by the USGS on earthquakes of the United States. A series of regional representatives each gave a talk about the earthquakes of their patch—starting with earthquakes of the East Coast, then earthquakes of the Midwest, and so on. Last up was a senior Alaskan seismologist. He started his talk something like this: "Well, in Alaska we have real earthquakes. Here's a map of California. Now this—[a line appeared on the map] is the rupture of the 1906 San Francisco earthquake. And this—transposed and at the same scale—is the rupture area of the 1964 Prince William Sound earthquake." And a rounded rectangle appeared on the map that covered most of the state.

Subduction zones are where the world's biggest earthquakes happen. In fact they are the only places where monster earthquakes can occur. It's a simple matter of geometry. The larger the area of fault that breaks, the bigger the earthquake you can get. Now in the case of the San Andreas Fault, you can have a fault break that is plenty long, but

Map of California with the rupture area of the 1964 Alaska earthquake plotted at the same scale. This earthquake was much larger than anything the San Andreas Fault could produce.

because the fault is vertical, its width is restricted to the depth of the crust, which is about twenty kilometers. So if the fault length was about 450 kilometers, its area is nine thousand square kilometers.

But in a subduction zone an oceanic plate is pushed under another (usually continental) plate. Instead of having a fault that is strictly vertical, this one slips downward at a shallow angle—usually starting at about 15 degrees and then gradually getting steeper. So the descending plate and the overlying plate rub together over a broad area—up to three hundred kilometers wide. In the case of the 1964 earthquake, where the fault break was about eight hundred kilometers long and three hundred kilometers wide, there was an area of more than 240,000 square kilometers of rock grinding against rock, which is vastly more than one can get in any other tectonic environment.

To get a really great earthquake, you need two things. It has to be in a subduction zone, and it must have a section sufficiently long and sufficiently straight that it can be one gigantic break. The coast of Chile is ideal—no surprises that this is where the world's record monster quake occurred in 1960.

But it is not the size of the largest subduction earthquakes that results in tsunamis. What is important is how they occur and, indeed, what happens when they are building up.

A typical subduction setup involves a fairly straight coastline—let's assume for convenience it runs north-south. Just offshore there will be an ocean trench, marking where the oceanic plate begins to bend downward as it is pushed beneath the continent. If you could see it all in cross section, you would see the oceanic plate, heavier and denser than the silica-rich rocks of the continental crust, sliding down at an angle that gradually gets steeper and steeper until it becomes a near-vertical plunge downward into the hot mantle, where the rocks largely melt and become absorbed into the mantle itself. (How do we know this? By plotting the depths of small earthquakes that accompany the whole process. They start shallow and get deeper and deeper until they die out about six hundred kilometers down.) If this could all happen smoothly, there would be no problem. But of course things get stuck. First, a piece of the plate moving downward gets stuck on the one above it. But the plate is still being pushed downward as new crust is created in midocean. The plate can't stop moving—it has to keep going. So it does. But the bit of overlying plate that is stuck to it now has to move with it.

Here is an easy demonstration. All you need is a flexible plastic ruler. Hold it firmly by one end in your left hand, so that it extends horizontally toward your right. Now press the tip of your right forefinger gently against the free end. Slowly move your finger diagonally downwards and to your left. The ruler is the continental plate. Your finger is the oceanic plate. Where you are pressing the tip is where the two plates have locked. What happens? Your finger drags the end of the ruler down and at the same time pushes it toward the other end of the ruler, so the ruler starts to bend and arch upward.

This is exactly what happens before a big subduction earthquake. The land nearest the ocean starts subsiding, while farther inland it starts to rise. If the locked area is out to sea, it may be that onshore you notice only the gradual upward movement, but the seabed nearest the fault is still going down. These movements are quite noticeable and have been recorded in Japan by teams of surveyors checking land altitudes. You can even detect them from satellite observations.

Eventually the elastic strength of the ruler is greater than the friction of the ruler tip against your finger, and the ruler springs back into shape—*per-doinggg!* That was an earthquake. But look—the tip of the ruler that was depressed flips up violently, and the bit that was bowed upward goes back to being straight.

This is more or less what happens in a subduction earthquake. The land wants to restore its proper shape, so the part that was subsiding flips up, and the part that was uplifted sinks back down. The trouble is that the part that flips up has a hundred thousand square kilometers of water sitting on top of it. And that's a lot of water to suddenly kick a few meters upward.

Imagine a bulge of water thrust violently up out of the ocean, several hundred kilometers long and as much as one hundred kilometers wide. Water being water, it can't stay in that position, so it immediately collapses and the water rushes away on both sides. Since we supposed the subduction fault was north-south, the bulge of water will be similarly aligned, and the water will form predominantly two waves, one going east and the other west.

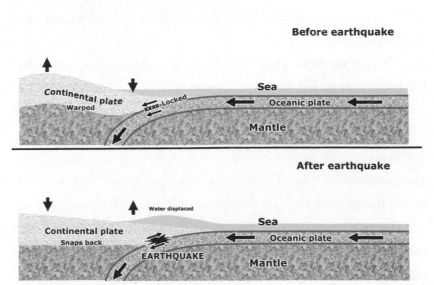

How great tsunamis are formed: where a subduction zone is locked by friction, tectonic movement drags the overlying plate downward. When the earthquake occurs, the overlying plate is released and rebounds upward, displacing a huge amount of water, which rushes away as a tsunami.

What happens next is controlled by the mechanics of water waves. An ordinary sea wave moves in a circular motion. If you could trace a particle of water, you would see that it moves in a circle, rotating in the same direction as the wave is moving—like the motion of a wheel. The bigger the circle, the faster the wave moves. What limits the size of the circle is the depth of the water. So in open ocean, waves can move as very large circles, and the distance between one wave peak and the next is accordingly large. So large, in fact, that it's hard to spot the waves at all. You can be traveling by boat in the open sea and have a major tsunami go right beneath you, and you may not notice it. It happens often. At this point the waves are traveling at about the same speed as a commercial airliner.

As the waves get closer to the shore, the water gets shallower. This limits the size of the circles of water, so the distance between wave peaks becomes shorter, and the waves slow down. But even though the waves in front slow down, the waves behind keep coming at the same speed, so the waves start to pile up. The water surges up the shore with tremendous force.

Now we can explain why Tilly Smith saw the sea going out before the tsunami arrived. Every wave has a peak and a trough. In the case of a tsunami, it's not just one peak and trough but a series of waves, like a succession of watery hills and valleys, that arrive one after the other. The leading wave, the one that arrives first, is usually a trough. As the seabed flips up, it thrusts the water above it upward; as that bulge collapses, the first movement is downward, so the first wave to spread out will also be a "down" wave, a trough. With respect to the level of a calm sea, wherever there is a peak, the sea surface is higher, and wherever there's a trough, the level is lower.

When the first waves arrive at the shore, the first peak slows down as the water piles up behind it. But the first trough doesn't need to slow down in the same way. All it does is lower the sea level. When the trough arrives, the sea disappears. And that is why the first sign of trouble is the sea mysteriously withdrawing from the shoreline, leaving fish flopping about on the now-dry seabed. It's as simple as the difference between a wave peak and a wave trough.

The distance between the first trough and the first peak can be so great that fifteen or twenty minutes may elapse between the first trough arriving on the beach and the first peak arriving behind it.

Even in places like Sumatra, not every great earthquake produces a great tsunami. Many thought the 8.7 magnitude earthquake that hit the Sumatra coast around Easter 2005 would result in another devastating tsunami like that of a few months earlier, just farther north. In fact the tsunami that occurred was small and caused little damage. It all depends on how much water is there to be displaced when the seabed jerks upward in an earthquake. While the 2005 quake certainly had a large fault movement, the sea above the displacement zone was shallow and surrounded several islands, so not much water was affected and a large coherent wave didn't develop.

The character of a tsunami depends on how much water is displaced, and how it happens. Imagine a one-ton concrete slab suspended by a crane three meters above a lake and then dropped in. How much of a splash does it make? It depends on how it enters the water. It makes a difference if the slab is suspended horizontally so the face of the slab smacks down onto the surface of the water, or if it is suspended vertically so that it enters the water edge-first. Even though the same amount of concrete is falling from the same height, the impact of the broader surface on the water maximizes the splash and the subsequent wave.

Alternatively, if a ton of gravel is dropped instead of a slab, the splash will be of a different sort entirely.

A subduction earthquake produces an effect similar to that of the concrete slab dropped on its face, except that the impact comes not from a slab dropped from above but from a slablike chunk of seabed that smacks upward from below. The result is a splash that produces a strong, coherent wave capable of traveling great distances.

In contrast, a tsunami produced by a landslide is more like tipping a load of gravel into the lake. The impact is not as violent as the smacking down of the slab, so the splash and wave are more fragmented. This is why the tsunami of Boxing Day 2004 was able to cross the Indian Ocean and still sweep up the beaches on the other side, whereas the

1929 Newfoundland tsunami was almost unnoticeable on the other side of the Atlantic.

This is also why Florida is not at risk, despite what some have suggested, from a supertsunami sweeping across the Atlantic from a collapsing volcano in the Canary Islands. It's true that the side of the Cumbre Vieja volcano has collapsed in the past and may do so again. But the sort of tsunami a collapse will produce will depend on how the collapsing mountainside enters the water. If a solid chunk of mountain were to tilt over and fall smack into the sea like a clown doing a pratfall, then, yes, that would produce a wave that might reach the other side of the Atlantic at a noticeable strength. But if it slides downward into the water, it will be like the underwater landslide of 1929: the tsunami will be strong locally but barely detectable at a distance. Needless to say, collapsing mountainsides don't generally flip over like a domino; they slide downward. Gravity sees to that.

This also applies to the supposed threat of a massive Pacific-wide tsunami from a volcanic collapse in Hawaii. There's no need to invent new dangers when the real threat of tsunamis from great earthquakes is bad enough.

SURFING THE TSUNAMI

One common misconception is that tsunamis look like huge break-ers, one of those big tube waves you see in surfing pictures. In par-ticular, it does not look at all like the famous picture by the Japanese artist Hokusai, *The Great Wave Off Kanagawa*. If I had a hundred dollars for every time that picture is used as the logo for a tsunami conference or workshop, journal, or website, I would be a wealthy man. But it's not even intended to be a tsunami—the artist just called it a great wave, which is what it is. At the time Hokusai was working (ca. 1831), no tsunami occurred that he could have used as a model.

One reason the myth of the giant tsunami wave has endured is the way tsunamis are reported. People read in the news that the earthquake produced a thirty-meter-high tsunami and suppose, not unreasonably, that this means a giant wall of water thirty meters from top to bottom. It would tower over most buildings as it swept toward them. But what this really means is that a tsunami got as far on land as thirty meters above sea level. A strong normal wave might sweep up to the back of the beach, which might be at an elevation of a meter or so, even while the wave itself had only a small crest.

So if a thirty-meter tsunami comes in, anyone who has run far enough to be thirty meters above sea level will be safe from anything worse than wet feet. The actual wave, from crest to trough, is rarely more than two meters high—still an impressive thing to see heading toward you at high speed.

The extensive videos of the two great tsunamis of recent years, Sumatra in 2004 and Japan in 2011, show what a tsunami really does look like—an evil, foaming, roiling mass of black water thick with debris and advancing relentlessly. Anyone trying to surf it would last a few seconds at most. Once the wave has started to sweep onshore, anyone in its way is likely to be killed outright.

As the tsunami picks up debris, its destructive power increases. It's bad enough for a building to have great masses of water rushing at it—but when cars, girders, beams, and boats are also smashed against it, the chances that the building will be destroyed increase greatly. A major tsunami really can pick up whole ships and toss them inland. After the 1960 tsunami in Chile, ships were found high and dry almost three kilometers from the coast. Not only that, but once the water has swept up to its farthest reach, it then flows back to sea again, carrying all its suspended junk with it. This scrapes away anything that survived the first passage of the wave.

An interesting case is that of Kamaishi, in Iwate prefecture, Japan. Kamaishi is a little fishing port on the east coast of Honshu, the Japanese mainland, and quite far north. The town is strung along a narrow valley that runs east-west between wooded hills, and there was a clear danger that any tsunami hitting the coast would be funneled up the valley with dire effect.

To protect the town, a breakwater was built at the mouth of the inlet. Given the depth of the water, this was a massive undertaking; construction took the best part of three decades and the cost was of the order of US$1.5 billion. It was finished only a couple of years before the great Tohoku earthquake of 2011 put it to the test.

Video footage shows exactly what happened. As soon as the earthquake shaking stopped, sirens started sounding around the town, warning inhabitants to take to high ground (not everyone paid attention). For a while, there was an eerie period of anticipation. The town had ridden out the earthquake quite successfully—hardly any damage to speak of, and this from a magnitude 9. Was that the end of it? What else was in store?

After about fourteen minutes observers looking out to sea noticed an unusual line of foam along the base of the breakwater. The tsunami had arrived; water was piling up on the seaward side of the breakwater and pouring over the top. The harbor basin began to fill up like a bathtub with both taps running.

Within five minutes the rising water had drowned the harbor piers and was lapping at the base of the cranes used for unloading the fishing boats. And still the water poured over the top of the breakwater in ever greater volume. Soon the breakwater was submerged and nothing stood in the way of the surging seawater, which rushed up the river estuary, sweeping away buildings effortlessly, as those watching from safety groaned at the horror of so much destruction.

The breakwater had not been built high enough. Few people had expected that such a large earthquake could occur, and those few who did were ignored. The earthquake catalog for the area showed no earthquakes greater than about 8.5, so this was the assumed limit.

But, again, before the 2004 Sumatra quake, few seismologists had ever seen a magnitude 9 in their working lives and were not really alert to seeing them in the historical record. In fact there was a precedent—an earthquake off the Honshu coast in 869. Although the records of it are sketchy, this was certainly substantially bigger than all the other quakes in the millennium that followed. The 869 quake should have warned people that magnitude 9 was not only possible but had already happened.

In many towns on the Japanese coast there was a well-defined "blue line" that was supposed to mark the maximum distance a tsunami could reach. People landward of the blue line comfortably assumed they had nothing to worry about. Not only was the seawall too low, the blue line was in the wrong place. What were supposed to be safe areas were not safe at all.

There is also a psychological phenomenon (known as *anchoring*) where people who have experienced something once tend to expect a second experience to be much the same. This affected many of the older residents along the northeast coast of Honshu who could recall the tsunami that reached Japan after the 1960 earthquake in Chile. They made the fatal mistake of thinking that all future tsunamis would be similar. But a tsunami arising from a giant earthquake on the other side of the planet and a tsunami from a giant earthquake only a short distance off your own coast are quite different. In many towns along the Honshu coast, because people never imagined that the water would ever top the levees that lined the estuaries, they wandered down to the riverside to watch.

Even when water started to pour over the levees, many did not appreciate the urgency of the situation until it was too late. One astonishing bit of video footage shows a man carrying a bulky television set up the street as he looks over his shoulder at the tsunami surging through the town toward him. He seems to have no inclination to drop the set and run for his life.

The breakwater at Kamaishi was not completely useless; it held the waves back for more than five minutes, giving people more time to get to high ground. Also, it put a dent in the force of the tsunami, and as a result it did not penetrate as far inland as it would otherwise have done. All the same, large areas of the seaward part of the town were razed by the force of the water, and about 1,250 people died—about 2.5 percent of the total population.

Farther south, the tsunami took forty-five minutes to reach the coastline at the Fukushima Daichi (Fukushima Number 1) nuclear power plant, its four boxlike main reactor buildings strung along the shore. Fukushima also had a defensive breakwater, designed to stop waves up to 5.7 meters. The wave on March 11, 2011, was fourteen meters. The water swept over the top of the breakwater and swirled up around the reactor buildings, flooding everything that water could get into. This included the emergency diesel generators that supplied the backup cooling system, which were essential, as the earthquake had shut off power from the main electricity grid.

Without a functioning cooling system, the situation in the reactors gradually deteriorated during the next few days, despite the battles fought by onsite engineers to bring the situation under control. Three of the four reactor buildings eventually exploded, with a major release of radioactive material into the environment. The Japanese government subsequently imposed a twenty-kilometer exclusion zone around the stricken plant; villages were evacuated, and the residents had no idea when, if ever, they would see their homes again. It was the worst nuclear accident anywhere in the world since the Chernobyl disaster in 1988.

The accident resulted from a combination of factors—the earthquake knocked out the primary power supply, the tsunami topped the breakwater, and the reserve power generators had been located where they could be flooded and put out of action. One lesson the nuclear industry has learned from the disaster is that it's not always the obvious things that will cause trouble. The obvious danger was always that strong shaking from an earthquake would damage the reactor or the reactor building, leading to the release of radioactivity. What was not so obvious was a chain reaction of incidents affecting the services that keep a plant running.

The Fukushima Daichi disaster prompted a reexamination of nuclear safety issues in several countries; in Germany it was largely responsible for a decision to close down the country's entire nuclear program, even though a tsunami is about as likely to hit Germany as it is to hit Oklahoma. Like Sumatra, this was a wave that changed things. The Lisbon earthquake and tsunami changed attitudes toward natural disasters and religion; the Tohoku event changed attitudes toward nuclear power.

And, one might add, it produced an indirect effect on climate change. When nuclear power stations are phased out, the generating capacity has to be taken up by conventional power stations, thus increasing carbon emissions.

Tsunamis can be avoided to some extent. I will discuss tsunami warning systems in Chapter 9, but tsunamis can be expected after large subduction earthquakes as a matter of course. Educational programs in places at risk from tsunamis, such as western Sumatra, try to condition people to react appropriately to earthquakes. The rule is that if you feel an earthquake and the shaking lasts a relatively long time, a tsunami is probably on the way. Therefore it's best to head for high ground or a designated tsunami shelter at once. If no tsunami arrives, or it's only a small one, not much time has been wasted. If a major tsunami comes, it may be a lifesaving move.

PART 2
SOLUTIONS

7
PREVENTION AND CURE

AS LONG AS EARTHQUAKES HAVE PRESENTED A DANGER TO human society, people have tried to lessen or avoid their effects. I call the four main strategies the four Ps: pray, predict, protect, and prevent. Pray that God will keep earthquakes away from you. Predict where and when the next earthquake will be, so that you can be sure to be somewhere else. Protect yourself so that when an earthquake does occur, it won't hurt you. Prevent earthquakes from happening altogether.

Prediction and protection are major topics that deserve two chapters each; the other two strategies need less discussion because by and large they don't work, so we can consider them first.

THE PATRON SAINT OF EARTHQUAKES

Prayer is, of course, the oldest strategy of all. Make enough offerings to Poseidon, and maybe he will be sufficiently mollified to not send any more quakes. In cultures that believed earthquakes were sent by God, asking God to stop them seemed a reasonable strategy. A common reaction to earthquakes in Europe during the Middle Ages and the Renaissance was to organize religious processions to demonstrate penitence, and this was not confined to cases of severe earthquakes; processions were sometimes triggered by only moderately strong shaking.

Many Christian societies interpreted moderate quakes as a warning from God that things were beginning to get out of hand. All the more reason to show piety and placate the Almighty before he got really angry and sent the big one. A major earthquake was the punishment, and contrition after one was a bit late. An earthquake in the Dover Straits in 1580 was strongly felt in both England and France, and many of the affected French towns immediately organized processions, which in one case continued as an annual event on the anniversary of the earthquake for many years afterward.[1] On the English side Protestants considered processions to be a Catholic thing. Instead, the government commissioned a special prayer book, and the order went out that all churches should read from it. Just over three hundred years later, when another damaging earthquake struck the east of England, one historically minded vicar dusted off a copy of the 1580 prayer book and used it once more.[2]

Religious reactions to earthquakes included giving thanks that things were not worse and that the destruction visited on a region was not total; God had mercifully spared the lives of the survivors, and this could be celebrated. Hence the need for solemn masses of celebration. A good example is the service organized in Naples following a strong earthquake in 1731—a special setting of the mass was commissioned from the composer Pergolesi, a native of Naples. Pergolesi emphasized the Gloria section of the mass to convey a positive message of God's majesty and power rather than giving the work a penitential character.

Once scientific explanations for earthquakes became the norm, blaming quakes on God's displeasure became rare. In 1863 a minor earthquake rattled the west of England, prompting the Anglican vicar of the small town of Leominster to preach a sermon blaming religious dissenters in the parish for the earthquake. Newspapers across the country made great fun of his antiquated and obsolete ideas.[3]

Even so, such attitudes can still be found today. In 2010 Hojjat ol-eslam Kazem Sediqi, a senior Iranian cleric, preached a sermon in which he stated that the only way to avoid being buried in rubble by earthquakes was "to take refuge in religion and to adapt our lives to Islam's moral codes."[4] Specifically he blamed women who wear revealing clothing as a cause of an increased incidence of earthquakes—perhaps not directly but by contributing to a general decline in moral standards. However, headlines in the West along the lines of "Scantily Dressed Women Cause Earthquakes" caused some ribald speculation amongst certain female seismologists as to how large a quake they could trigger if they dressed down a bit.

There is always someone you can blame. If an earthquake strikes a region where the predominant religion is not yours, you can say that God is smiting the unbelievers. If it strikes your own coreligionists, you can claim that God sent the earthquake because people were not devout enough. This particular line of argument seems endless, since there will always be some who don't meet the strictest standards of devotion. Any group can point to any other group and say, "This is all because God is angry with YOU!" It seems rather stultifying.

Unfortunately, there is a very real danger that religious fatalism will interfere with practical efforts to protect communities from earthquakes. Sediqi's message was, in effect, that human measures to reduce earthquake risk are pointless, and one should look only to improving the country's morals.

If you do want to put your trust in prayer, there is a saint to pray to—Saint Emidius, patron saint of protection from earthquakes. According to legend, he was a third-century bishop who started life as a pagan in Trier, Germany. After converting to Christianity, he traveled to Rome, where he upset many by smashing pagan idols. Promoted to bishop of Ascoli Piceno in the Marche region of Italy, he was subsequently beheaded around 309 by the local governor. As with so many early saints, there is considerable doubt as to whether he really existed.

His association with earthquakes has nothing whatsoever to do with what he did during his lifetime. Because of his local association with Ascoli, he became the town's patron saint, and for many people, if they know anything about St Emidius, it will be through the famous painting of him (completed in 1486) by Carlo Crivelli, which depicts Emidius witnessing the Annunciation while holding a model of the town of Ascoli. (The painting is in the National Gallery, London.) In 1703 an earthquake caused considerable destruction in the Marche region, but damage was much lighter in Ascoli. This was attributed to the protection of the local saint, and Saint Emidius then had "protection against earthquakes" added to his responsibilities.[5]

This was further confirmed after the 1731 earthquake in Naples. Pergolesi dedicated his mass to Saint Emidius, and the city of Naples decided to adopt him as its patron as well.

While prayer is an option, there is not much evidence to suggest that it has been a particularly effective one. Being devout didn't help the people who perished in the ruins of the churches of Lisbon in 1755.

PUMPING THE FAULTS

If earthquakes happen only because of loose morals, then, in theory, if everyone adopted a suitably religious lifestyle, earthquakes would stop. I have not yet heard anyone expound on exactly how this would happen in geological terms. Would Earth's tectonic plates suddenly grind to a halt? What other consequences would there be? I suspect that even Imam Sediqi has not concerned himself to think this out.

But other than placating a wrathful God, is there any other way to prevent earthquakes? I sometimes joke that we need to hammer in giant nails to fix the tectonic plates in place. Even if this wasn't ludicrous, it wouldn't work, because the forces that cause plate movement would continue to operate, and stresses would continue to build up until something eventually let rip.

To stop the whole process one would have to tackle the problem at its root and cut off the power supply. Since this would mean somehow draining off heat from deep inside Earth's mantle, this is beyond even the most fanciful daydreams.

But there is another line of approach. We can't stop faults from moving—but maybe we could control the way they move. After all, some faults slip smoothly in a seismic creep. The problem is not the fault movement but the stick-slip process. If it were possible to somehow lubricate a fault so that it always moved smoothly, earthquakes would not occur. So long as you didn't build a house directly across the fault trace (where it would be gradually torn apart), the fault would not be dangerous.

Modifying fault behavior had its most popular expression in the 1985 James Bond film, *A View to a Kill*, in which the villainous Max Zorin plots to destroy California's Silicon Valley by triggering a massive earthquake. Zorin's idea was to stuff a mine full of explosives at a point where the San Andreas Fault is locked, so that the blast would break the fault loose and cause an earthquake. Subsidiary explosions under lakes along both the San Andreas and Hayward Faults would flood the faults and allow them to slip more easily.

It's easy to see why this would probably fail. The points of high friction (known as asperities) locking the San Andreas Fault are at least five kilometers deep—how deep is that mine? No explosion Zorin can create is going to scoop out a crater large enough to break the asperity where it matters. The best one could hope for would be that the transient energy wave from the explosion would give the fault enough of a kick to set it moving—if it was close to failing already. And it is hard to see that those explosions under the lakes are going to do much more than blast water upward and kill a lot of fish.

To make sure water goes down into the fault, you need a pump, not an explosion. And this has been shown to work, at least on a small scale. If you pump water deep into the ground, it changes the water pressure within pores in the rocks, and this has a lubricating effect on faults. The overall effect is that it requires less stress for the fault to slip. But this does not mean that the fault becomes smooth-slipping; it just produces earthquakes more easily.

So while it may not be possible to turn a locked fault into a smooth-slipping fault, in theory fault behavior could be controlled through a series of small earthquakes that would preempt the occurrence of a large one.

To stay with Northern California as an example: the magnitude of the 1906 San Francisco quake was just below 8, but let us round it up to 8 for the sake of argument. At some point one can expect a recurrence—a similarly sized earthquake that will break more or less the same length of fault as ruptured in 1906. Current opinion among seismologists is that such a repeat earthquake may be due around 2030—but don't put this on your calendar, for the uncertainties are large.

One could argue along this line: In about twenty years' time the strain that has built up since 1906 will be released in a magnitude 8 earthquake that will do a lot of damage. To stop that, we need to dissipate that strain energy in advance. Smaller earthquakes are less damaging than large ones, so if we can release the energy in a series of controlled smaller earthquakes, we can prevent the large earthquake from ever happening.

In theory this might be possible, although it's hard to know exactly what would happen if engineers just started pumping water into bits of the San Andreas. But let us suppose that problem is solved. Let us suppose a way could be found to know in advance that if you pumped X million liters exactly here, here, and here, you could say precisely what magnitude earthquake would happen as a result—would that solve the problem?

The crunch is the way energy translates into magnitude, as I discussed earlier (in Chapter 5). To use up all the energy that makes a magnitude 8 earthquake, you would need the equivalent of thirty magnitude 7 earthquakes. Remember that the Haiti earthquake was a 7; that is how much shaking a 7 can cause. Do you really want to face thirty of those in Northern California during the next twenty years? So let's opt for magnitude 6s instead. Now we need nine hundred of them to do the job, strung out across the 450 kilometers or so from Shelter Cove to San Juan Bautista. That's a lot of earthquakes—and even a magnitude 6 can cause damage. With a target of nine hundred quakes in twenty years, one would need roughly one a week to dissipate all that energy.

Would people tolerate that? Furthermore, suppose one of those triggered quakes damaged your house or, worse, killed someone? Who would be responsible? Acrimonious law cases occurred recently in Switzerland concerning an earthquake of magnitude 3.5 that was accidentally triggered by water pumping near the city of Basel as part of a geothermal scheme. What would be the legal repercussions of a magnitude 6 earthquake that was deliberately triggered in California? One would need to have laws in place that specifically gave legal exemption from responsibility for any damage to whoever was in charge of the earthquake control program. Otherwise even trial runs for such a program would be a legal nightmare.

So regardless of whether it is possible to solve the technical problems for earthquake control, the social, legal, and political issues are so daunting that no one is harboring any dreams of trying it any time soon.

With prayer doubtful and prevention impractical, let's look at the other two options: prediction and protection. What can they offer?

8
NEXT YEAR'S EARTHQUAKES

EARTHQUAKE PREDICTION IS SOMETHING THAT NO SEIS-
mologist can get away from. After any major earthquake, usually the
second question from reporters (after "Why did it happen?") is "Was
the earthquake predicted?" If not, then the journalist asks either, "Can
you predict earthquakes?" or, "Why can't you predict earthquakes?"

Here are two of several answers to the last of these questions.

Compare earthquakes to the weather. We can make weather fore-
casts, but they are often wrong. Weather is something we can see. We
can see clouds gathering and know that a depression is coming our way.
These days we can even see from space everything that's going on in the
atmosphere. We can track every weather system, every cloud. Yet we are
still only so-so at predicting the weather. Now imagine that all the clouds,
all the weather systems, are miles underground, out of sight. Want to try
making a weather forecast now? Well, that's how it is with earthquakes.

Now imagine again you are holding a plastic ruler in both hands
and slowly bending it. Keep slowly applying pressure until it snaps. As
you watch it bending, can you say exactly when it will snap? An earthquake fault is much the same: strain builds up until it breaks, but when
is that exact moment it finally snaps?

Nevertheless that doesn't stop people from trying to find a way.
Back in the 1960s it was routinely assumed that earthquake prediction

was just around the corner, that in a couple of decades there would be bulletins of next year's expected earthquakes as a matter of course. It didn't turn out that way. Fifty years later we are still no closer to a working earthquake prediction service. Meanwhile an army of amateurs has stepped up to try their hand where the professionals have failed.

There's a well-known saying that it's difficult to make predictions, especially about the future.[1] This is certainly true about earthquakes.

The problem that has bedeviled earthquake prediction is that so many ideas have turned out to be dead ends. However, no one wins a prize for finding something doesn't work. Aspiring scientists are always told that an honest negative result is still a useful thing; it stops others from wasting their time by going down the same path. But it's neither glamorous nor attention-getting. The result is a tendency to oversell new ideas instead of assessing them critically and objectively. So other seismologists have had to take on the role of evaluators, and even debunkers, for every new earthquake prediction theory. As a result, earthquake prediction is perhaps the most controversial, and adversarial, field of seismological research.

The way that science ought to work is that any new method or hypothesis is presumed guilty until proved innocent—or, rather, false until proved true. (To be strictly correct, I shouldn't say "proved true"— shown not to be disprovable would be a better way of putting it.) So if a method for earthquake prediction is proposed, the first assumption ought to be that it doesn't work. And one should set out to test that. The technical term used is the *null hypothesis. Null* means *nothing*—the null hypothesis represents the status quo. For anything you study in science, you start with the null hypothesis that what you are looking at is a chance effect, it doesn't mean anything, the method doesn't work, there's not really any signal, or whatever. Only when the data are such that the null hypothesis is shown to be untenable are you allowed to think you might have found something useful.

But let's start with some basics. How could one begin to approach the problem of earthquake prediction? There are three main sorts of method one can apply.

The first is pattern recognition. This is definitely the easiest approach, as it requires no equipment and no experimentation, just a lot of historical data and a good imagination. The essence of the method is to trawl through past records in search of repeating patterns, so that one can say things like, "If there was an earthquake at X last year and an earthquake at Y this year, then next year there will surely be an earthquake at Z." This is an ideal pastime for the armchair seismologist.

The second approach is the precursor method. Here one assumes that the process by which a fault fails produces not only an earthquake but also some sort of other detectable phenomenon in the days before the earthquake actually happens. It might produce electrical signals. It might emit streams of ions that heat the upper atmosphere. It might expel gases such as radioactive radon. It might do something to groundwater levels. It might do something that disturbs animals, causing them to behave oddly. All these have been suggested. And it certainly might produce small preparatory earthquakes—foreshocks.

The third approach is slightly different. It assumes that even if the fault failure process does nothing detectable, a fault will fail only when highly stressed, and increased levels of stress may change the properties of rocks in the vicinity in ways that are measurable. Notably, some seismologists have suggested that rocks behave like a sponge does when squeezed. Just as the pores of a sponge close up under pressure, so do tiny cracks in rocks. When this happens, the cracks, like the closed pores of a sponge, will be aligned at right angles to the direction of pressure. This has an interesting effect: it makes the rocks polarizing, like the lenses of expensive sunglasses. Just as polarizing sunglasses affect the way light travels through them, rocks with aligned cracks affect the way seismic waves (in particular, shear waves) travel through them. By tracking how shear waves are affected, in theory a seismologist can draw conclusions about the state of the stress in the surrounding rocks. While it seems logical in principle, the sort of monitoring required is difficult and expensive to do, and the method is largely untried.[2]

A SEISMOLOGICAL HEADACHE

You generally don't get amateurs having a go at scientific problems in areas like particle physics, since not many could afford the necessary lab facilities. But earthquake prediction looks superficially like a good opportunity for the armchair scientist equipped with a few files of data, a good imagination, and a capacity for self-delusion. As a result entire sites on the Internet are devoted to the efforts of amateur earthquake predictors, and the atmosphere can sometimes be poisonous as rivals dispute each other's track record.

Perhaps the two most popular approaches are those based on planetary alignments and *earthquake sensitives*.

That earthquakes might be influenced by the gravitational pull of the sun and moon, and perhaps more distant bodies in the solar system, is an old idea. It has two basic problems. First, let's suppose that gravitational forces from the sun and moon do have an effect on earthquake activity. What those forces do not do is create the energy necessary for earthquakes. Seismic activity is the result of tectonic forces produced as the plates shuffle around, driven by movement in the mantle. Gravitational pull can't produce an earthquake out of nowhere. The best gravitational pull could do is exert a weak force that could be the final trigger for a fault that was already poised to go—advancing the time of the next earthquake by maybe a few days or weeks.

This is not very useful for predictions. You can have all the planetary alignments you like, but if a fault is not primed and ready, it won't budge.

The second problem is that, given the large number of earthquakes that happen every day, it's hard to know if gravitational pull is triggering any of them. It's no good saying, "I saw there was a conjunction of Mars and Jupiter, so I thought there would be an earthquake, and there was." Earthquakes are always going on somewhere, no matter what Mars and Jupiter are up to. Triggering is at best a weak effect and not of much practical use, even if it were possible to prove.

So if ever you see an article claiming "XXX may trigger earthquakes," remember that *trigger* does not mean *cause*. A useful analogy is an old-fashioned spring-loaded mousetrap. If it is primed and set, a small nudge may trigger it. If it's not primed, you can kick it round the kitchen and not set it off. So unless you know a fault is about to break (and you never do), information about a potential trigger is of no use.

The other predominant class of amateurs, earthquake sensitives, are an interesting crowd. These are people who claim to have a special sensitivity that allows them to detect earthquake precursors that are not detectable in any other way. The signals they detect manifest themselves either as a pain in some part of the body or as a heard sound, typically a sort of whistling.

On the face of it these claims seem implausible. It is hard to understand how, say, an impending earthquake in New Zealand could communicate a signal, undetectable by any sort of instrument, to a person in Chicago. In some cases, it is clear that the individual concerned has simply made an unjustified connection between different phenomena, such as having an unusual headache and then hearing an earthquake reported on the news a few days later. People who are unaware of the actual frequency of earthquakes might convince themselves they are sensitive, not realizing that if this were true, they would probably have headaches constantly.

Where predictions based on earthquake sensitivity have been tested, the result is usually insignificant, though this is not always reported honestly. I looked once at the website of a sensitive who claimed to have had her claims tested and vindicated by the USGS. I took the trouble to check this, and the USGS seismologist involved told quite a different story.

Despite such cases, some earthquake sensitives are honest, intelligent, and sincere and believe they have a real gift, even if they can't explain it. Even if such gifts exist, though, it still doesn't get us very far. An inexplicable personal method of predicting earthquakes has to be filed under paranormal phenomena, not seismology. I tell people that I'm really not interested in whether they can predict earthquakes. If they can teach me to predict earthquakes, then I'm interested. Science has to be transmissible, or it's not science.

CLUMPS AND CLUSTERS

Looking at seismology from an international perspective, it is sometimes possible to detect national schools of thought, presumably through methods and approaches handed down from professor to student at major universities. Russian seismology seems to have a tradition of always developing the most mathematically complex approach to any problem. If you go to the hall at an international seismological conference where contributions are pinned up as posters, and you see in the distance a board with a dozen sheets of paper dense with typewritten equations, chances are the authors are Russian. Pattern recognition lends itself especially well to advanced mathematical modeling, so it's not surprising that seismologists who champion this approach are often either Russian or trained in Russia.

But for the rest of the seismological community, the question is whether all those equations actually reveal what is going on inside the earth.

The first assumption has to be that null hypothesis—in this case that earthquakes are happening semirandomly within certain fault systems, and any patterns you might think you see are just chance. Even if you can create a set of equations that reproduces the pattern that happened in the past, what happens in the future might be quite different.

It's like taking a handful of pebbles and throwing them on the sand. If you look closely at the result, you'll see that some pebbles have fallen in a straight line; elsewhere you might spot a ring and some triangle patterns. But pick them up and throw them again, and the straight lines are somewhere else, and there's no ring anymore.

The best-known pattern recognition approach is the M8 algorithm, developed by Vladimir Keilis-Borok, formerly of the Academy of Sciences in Moscow and now at the University of California. His approach is based on a series of sliding averages of past earthquake activity, from which various parameters are computed using a number of mathematical functions. When several of these functions together yield larger-than-expected values, it's considered a danger sign.[3]

The danger sign is translated into a "time of increased probability," or TIP, which typically lasts for five years. This is not exactly what most people think of when they think of prediction. An earthquake prediction, if one is going to be precise about it, requires three things: when, where, and how big. Ideally, a prediction would say something along the lines of "between September 14 and 18 next year, there will be an earthquake in the range of 7.0 to 7.5 within fifty kilometers of Antofagasta, Chile." To say that within the next five years an earthquake is more likely than usual to occur in an area covering hundreds of square miles is much weaker and perhaps not very useful. Indeed, one can draw a distinction between a prediction (which says an earthquake will occur within narrowly defined bounds) and a forecast (which says it might, within vaguely specified bounds), in which case M8 is more a forecasting tool than a prediction method.

Keilis-Borok and his team ran the M8 algorithm retrospectively to see if past large earthquakes corresponded to TIPs, with good results. But alas, prediction about the future is more difficult. The track record of M8 in forecasting future earthquakes has not been inspiring.[4]

There are two big stories about pattern recognition predictions. Sadly, both are failures.

The first dates to the 1970s. Brian Brady, an American seismologist, had begun looking at rock bursts in mines (these are something like a cross between an earthquake and an explosion—rock shatters violently under extreme pressure; such bursts are a major hazard to miners). Brady concluded he could recognize typical patterns of behavior that preceded a major rock burst. More controversially, he also believed he could scale this up to the level of earthquakes.[5]

Brady started examining the record of earthquakes in California, looking for clues that suggested that similar patterns were occurring, and he believed he had found them. When a great earthquake of magnitude 8.1 struck off the coast of Peru in 1974, his attention traveled south. It seemed to Brady that he was seeing the preparation phase for a still greater earthquake, one that would rival the 1960 Chilean earthquake in size, and perhaps exceed it. It would perhaps be the biggest earthquake in history. He set a date: mid to late 1981.

Not surprisingly, this alarmed the Peruvians, as word of the prediction began to leak out to the general public—particularly when casualty estimates ran to five figures. Whether it would be the world's largest quake or not, it looked like it would be South America's deadliest. But was it really going to happen?

A lot of people decided they were not going to take any chances. Investment in Peru slumped, property prices fell, tourists stayed away. I remember that even in faraway Edinburgh, I was being asked if it was safe to travel to Peru.

Making a rational assessment of Brady's prediction was difficult. His published description of his scientific basis was obscure, conceptual, incomplete, hard to follow, and downright weird. By January 1981 the Peruvian government appealed to the United States for advice on how to proceed.

In response, an expert panel of nine senior seismologists was assembled to assess the reliability or otherwise of the prediction. Brady and his coauthor, William Spence (who played a relatively minor part in the development of the prediction), were given five hours to explain the basis of it all.

Brady was aggrieved; he had counted on a longer time to take the panel step by step though his ideas. Pressed, he began to ramble through more and more strange, speculative ideas. The panel members were baffled. They had expected a clearly reasoned, seismological case, backed up by hard evidence. What they got was an elaborate mishmash, ranging even into the field of cosmology.

After a second day of largely fruitless cross-examination, the panel found themselves completely unable to understand Brady's hypotheses on which the prediction was based. Either Brady was a genius whose ideas were so lofty all nine experts could not even perceive the foothills—or it was just so much fantasy. The first option didn't seem likely; surely at least one of the nine would have a glimmer of understanding. They settled for the second. Their verdict was that the prediction should be discounted. Even so, a lot of people in Peru were spooked by the whole thing and unsure if they were really off the hook or not.

Brady was now changing the details of the prediction, both dates and magnitudes. The first foreshock, he now said, would come on June 28. By the time June rolled around, even his colleague Spence had lost faith. The absence of any earthquake on June 28 was front-page news in Peru. "Nothing Happens" crowed the headline in one Lima paper.

By July 9 even Brady had lost faith, and later that month he formally withdrew his prediction. The Peruvian earthquake prediction scare was over—but the economic consequences lingered on for years. There may have been no earthquake disaster, but the prediction itself was an economic disaster—a man-made one—for Peru.

LIKE CLOCKWORK

The Brady prediction is one cautionary tale; the second, fortunately, had no such economic consequences, and the worst effects were some red faces. It concerns a little place called Parkfield, California.

All seismologists know exactly where Parkfield is, but it's virtually unknown to the general public outside its twenty or so inhabitants. It's not exactly a bustling metropolis. Well, actually, it does host a mountain bike race every October, so some nonseismologists do know about it, but it's not so much small-town America as tiny-village America. What it's remarkable for, though, is earthquakes. The Parkfield Café, should you walk in for a coffee, proclaims Parkfield as "Earthquake Capital of the World."

The village lies on the San Andreas Fault. Now, the San Andreas is very long, reaching right down the length of California from north to south, and it doesn't all behave in the same way. Some bits, for instance, the Fort Tejon sector north of Los Angeles, seem to stick shut for ages and then break in one big earthquake. Other portions appear to slip smoothly without any earthquakes at all. But the bit near Parkfield appears to have a characteristic pattern of breaking whenever it has enough strain accumulated for a magnitude 5.5 to 6 earthquake— moderate; not great and not small.

The record certainly looks consistent. The first traceable Parkfield earthquake was back in 1857; then came similar quakes in 1881, 1901, 1922, 1934, and 1966. Given how ill-behaved earthquakes usually are, this looks pretty regular: intervals of 24, 20, 21, 12, 32 years—average 21.8; call it 22. When this was spotted in the early 1980s, it wasn't too hard to add 22 to 1966 and come up with 1988 as the expected date of the next Parkfield quake. However, since the gap wasn't always exactly 22 years, it was necessary to put some uncertainty on that figure, and it was confidently asserted that the next Parkfield quake would be between 1985 and 1993 with 95 percent probability.[6]

1988 came and went. No earthquake. 1993 came and went. No earthquake. Eventually in 1996 Yan Kagan at the University of California showed that statistically the analysis that yielded the 1985–93 prediction was seriously flawed.[7] The 1966 plus 22 calculation was just not dependable, and it was no surprise the expected earthquake had not materialized. Earthquakes don't run like clockwork, not even in Parkfield.

Eventually the earthquake occurred—on September 28, 2004, sixteen years late. But this earthquake turned out to be important for earthquake prediction in other ways. As Comrie was in the 1840s, so was Parkfield in the late twentieth century—a natural laboratory for earthquakes, specifically, for earthquake prediction.

Since it was expected that a significant earthquake was going to happen soon on the Parkfield segment, this was an ideal opportunity to look for something detectable that precedes an earthquake. Parkfield became one of the most intensively monitored bits of real estate anywhere. Seismologists looked for every sort of potential precursor. Magnetometers, gravity meters, tiltmeters, strain meters . . . if a mouse sneezed, it would be detected.

I'll come back to that.

WATCHING THE YAKS

Speaking of mice, many people believe animal behavior is a potential precursor of earthquakes.

A lot of tales about animal behavior are squarely in the realm of folklore. Even in cases of seismically quiet countries, you will hear stories about how Tiddles the cat was acting strangely before the earthquake struck. This is frequently a variation on a well-known logical fallacy, the technical name of which is *post hoc ergo propter hoc*—after this, therefore because of this. "I prayed to God for there to be no earthquake; there was no earthquake; therefore my prayers prevented the earthquake." In this case we have, "Before this, therefore because of this," which is even less likely. "Tiddles was uneasy; then there was an earthquake; therefore Tiddles was uneasy because of the earthquake." That's not really logical. If an earthquake occurs, you may remember that Tiddles was uneasy the day before. If nothing untoward happened, you would just forget about it. Cats are cats.

Scientifically testing animal behavior during earthquakes takes a lot of effort, but because of the sheer amount of anecdotal material, some scientists have taken this sort of research seriously. One Japanese investigator, Tsuneji Rikitake, attempted to assemble all the accounts he could find of strange animal behaviors before earthquakes.[8] Some of it is vaguely amusing, as in "yaks in the zoo refused to eat their food."

Another long-running experiment in Japan involved catfish. Japanese mythology explains earthquakes as the movement of giant catfish under the sea, so just in case there was something in this, some catfish were put in a laboratory tank under constant surveillance to see if they behaved in any special way before an earthquake occurred. They never did.

Other laboratory experiments have included keeping small animals in one corner of the lab and a rock crusher in the other. This is a giant press that squeezes rock samples under a huge piston. In theory the rock crusher can create stress conditions similar to those before an earthquake, and scientists carefully observe how the animals react. At Osaka University in Japan, Motoji Ikeya and colleagues squeezed blocks of granite in a five-hundred-ton compressor until they broke, while keeping budgerigars, rats, mice, eels, and silkworms in tanks or cages nearby. Just before the rock finally broke, the budgerigars started twittering, the mice looked alarmed, and the rats woke up and started looking around and washing their faces nervously.[9] It's a fascinating result, but does it really correspond to what happens before real earthquakes? Or were the animals reacting to a change in the noise of the machine?

The most recent animal to be featured in the annals of prediction is the toad, after a colony of the creatures was observed abandoning its pond near L'Aquila, Italy, shortly before the 2009 earthquake there.[10]

Various ideas have been put forward to explain why animals might be sensitive to impending earthquakes. If a domestic pet reacts immediately before a quake strikes—just seconds before—it may be that, being more sensitive than humans, it has noticed a weak P-wave arriving before the stronger S-wave starts shaking the furniture. That seems reasonable—but if it behaves oddly days in advance, if it is not to be written off as coincidence, then one has to suppose that the animal has detected something that no instrument has picked up.

One suggestion is radon gas, squeezed out of the highly stressed rocks, which is sometimes detected instrumentally before earthquakes. In the case of the Italian toads, the supposition is that charged particles emitted from the ground changed the character of the pond water. The toads, who are sensitive to the water chemistry, didn't like the change, and departed.

BIG QUAKES AND LITTLE QUAKES

The easiest and most obvious precursor is certainly foreshocks. Just as aftershocks are the smaller earthquakes that usually arrive in the wake of a large earthquake, small quakes can occur before the main shock. So with the benefit of hindsight, you can look at an earthquake sequence and say, here are the foreshocks—then the main shock—then the aftershocks. So why not watch for foreshocks to determine when a large earthquake is on the way?

Once again, it's difficult to make predictions, especially about the future. The problem with foreshocks is that they look exactly like any other small earthquake. What marks them out as foreshocks is the fact that a big main shock happens afterwards. Until the main shock occurs, they seem just like any other small earthquake. And as small earthquakes are common, if a big earthquake were predicted every time a small earthquake occurred, people would get tired of it after the second or third time. The other problem is that major earthquakes often have no foreshocks at all.

Nevertheless, sometimes one can gain something by recognizing particular patterns. If a historical major earthquake in a certain location was preceded by many foreshocks and no earthquakes have occurred in that location since, and then small earthquakes suddenly started appearing again, it would be reasonable to worry. There is a classic foreshock pattern in which, after decades of silence, small quakes start happening, build up in frequency, and suddenly stop. The main shock then follows within twenty-four hours or so.

That's useful for predicting earthquakes if it can be spotted and recognized, but you certainly can't bank on it. In the Bam case, which I mentioned in Chapter 1, for instance, it is said that foreshocks were mistaken by the local inhabitants for something else entirely and were too small to be picked up by seismologists.

An entirely different type of problem with foreshocks arose with the destructive L'Aquila earthquake of 2009 in central Italy. Here, a significant foreshock sequence was recorded by seismologists, but they disregarded it. They did so for the good reason that Italian earthquakes generally don't have foreshock sequences, and sequences of small ordinary earthquakes are common. So if every sequence of minor tremors was identified as a potential foreshock sequence, the seismologists would usually be wrong, and crying wolf so often would not help matters at all.

Additionally in the L'Aquila case, a local lab technician made a very public earthquake prediction (he drove around in a car with a loud hailer to broadcast it) based on the detection of radon gas. Despite the fact that an earthquake did occur, his prediction was unsound and wrong in the details. Not only did his prediction not have a good technical basis, if the local inhabitants had taken his advice, they would have evacuated safe areas and moved into dangerous ones before the quake struck. But the inhabitants of L'Aquila were naturally concerned. A commission of seven experts was convened to give advice, and they downplayed the threat.

Then, on April 6, disaster struck. A magnitude 6.3 earthquake hit the ancient city in the middle of the night. Buildings throughout the city were damaged, some collapsing. Many of the old medieval churches were particularly badly damaged. But modern buildings often did no better, or fared even worse. News reports highlighted the case of a dormitory at the local university that collapsed around the heads of the sleeping students. Even some buildings that were supposed to be earthquake-resistant fared badly, suggesting they had not been built well at all. In total, 308 were killed, the worst death toll in Italy since 1980. Damage was also heavy in some of the outlying villages, a few of which were totally ruined.

After the earthquake, the members of the commission were subsequently indicted for manslaughter, for failing to give sufficient indication of the seriousness of the situation. This caused outrage in the scientific community, partly because it was reported that the experts were prosecuted for failing to predict the earthquake (none of the seven was even an expert in prediction)—obviously a quite unreasonable charge. The reality is a little more subtle. The considered opinion given out by the commission was that it was impossible to say whether the minor quakes were foreshocks or not, and that the balance of probability was that they were not, since that was the more common experience in Italy. This was a statement that no seismologist would disagree with.

However, the seventh member of the commission, the only one who was not a seismologist or earthquake engineer (he was an official in the civil protection department), made a statement that the occurrence of minor quakes made a larger one less likely, since it showed that energy was being released harmlessly. This is something that no seismologist would endorse. The charge is that this pronouncement gave local inhabitants a false sense of security and dissuaded them from fleeing the town. The charge against the other six appears to be that they didn't jump on this statement as unjustified. At the time of writing, it's unclear how this case will play out. But scientists are watching developments with unease, uncomfortable with the idea that scientific opinion given in good faith might land one in a court of law on criminal charges.

There has been one major success story in prediction, involving foreshock activity: Haicheng, in northeastern China. After a series of minor earthquakes in 1975, seismologists decided that something larger was coming and issued an alarm. Evacuations were conducted, despite the cold winter weather, and when a major earthquake struck, many of the buildings that were damaged were empty.[11]

Any sense of jubilation in China that earthquake prediction could now be relied on proved short-lived. The very next year the Tangshan earthquake struck, taking a quarter of a million lives with no warning and no prediction.

And this is the problem with using precursors such as foreshocks to predict major earthquakes: the utter lack of reliability. Very occasionally an earthquake will telegraph its punch, as in Haicheng, but what works with one earthquake fails with the next. It's like having an armor vest that stops every second bullet—it's the ones it doesn't stop that kill you.

It's much the same with the emission of radon gas from the ground before an earthquake and, for that matter, odd animal behavior. Sometimes it's observed. Other times it isn't.

Essentially any type of precursor yields four possibilities:

Precursor observed, earthquake happens.
No precursor, earthquake happens.
Precursor observed, no earthquake.
No precursor, no earthquake.

For a system to be really useful, the only possibilities should be the first and the fourth. If an earthquake occurs with no precursor, you can get a disaster like Tangshan. If there's a precursor but no earthquake, it can be highly inconvenient and expensive (as witness Peru)—and there were reportedly quite a few cases in China in the 1970s where evacuations were ordered for earthquakes that never happened.

A prediction system in which either an earthquake infallibly follows a precursor, or if there is no precursor there is no earthquake, would be a perfect system. But the prediction community often complains that critics disregard any fledgling system that isn't perfect, and, well, Rome wasn't built in a day. The problem is how to distinguish a system that is useless from one that it is imperfect but promising.

KEEPING SCORE

And this brings us to the world of prediction evaluation, and a murky world it is, riven with controversies and disputes. It should be straightforward: count up the number of successes and failures. If there are plenty of successes, then things are looking good. The trouble is that there are so many ways to abuse the system, and not everyone is pure-mindedly seeking the advancement of science.

Despite the difficulty of making predictions about the future, there are some predictions that just can't fail. I'll make one now: there will be an earthquake tomorrow. Well, of course there will be; dozens of earthquakes are recorded every day, all around the world. So this is a true prediction but a useless one. However, suppose there is a major earthquake tomorrow? I could say, "Well, I said yesterday that there would be an earthquake today, and here it is"—and how many nonseismologists would detect that I was exaggerating my powers of prescience?

Technically that wasn't even a prediction—by seismological definition a prediction must state place, time, and magnitude, and I stated only time (tomorrow). But I could still make a surefire prediction by specifying the other two with wide ranges—tomorrow there will be an earthquake with magnitude greater than 4 in the Pacific region. I guarantee it. I could even be more specific and still have a huge chance of success—magnitude greater than 6, Pacific region, next week. It's likely to happen whichever week you pick.

If you really want to get your name in the newspaper as an earthquake predictor, here's a sneaky way to do it that only requires an accomplice. Pick a few likely locations for earthquakes—let's say, California,

Chile, and Japan. At the start of the year, send some emails to your accomplice. You need twelve separate emails for each location. The first will read, "This January, there will be a major (7 or larger) earthquake in California." The next one, "This February, there will be a major (7 or larger) earthquake in California." And so on, and so on. If it turns out that there is a major earthquake, say, in Chile in July, your accomplice digs out the email that says, "This July, there will be a major (7 or larger) earthquake in Chile" and sends it off to the media as proof of your prediction prowess. Since the email is date-stamped at the beginning of January, you can prove that you made the prediction well in advance, and of course no one will ever know about all the other emails, which will be quietly deleted.

Demonstrating how you made the prediction will be difficult, but the point is, there are ways of cheating, and snake-oil merchants are not unknown.

Another way of getting yourself an undeserved reputation for predicting earthquakes revolves around the concept of "better than random." The argument here goes, "No one can expect my predictions to be perfect. But in tests they have been shown to be significantly better than random." This sounds good. It sounds like a proper scientific evaluation has been carried out and returned a positive result. In fact it can easily be faked in a way that the unwary will not notice.

Draw up a list of predictions for next year. California—magnitude 4.5 to 5.0 within five days of February 22. Chile—magnitude 5.0 to 6.0 within five days of March 15. And so on. These are sufficiently likely events that, with any luck, some will come true.

The next step is to create a control set. These will be the same number of predictions, but this time they will be completely random. Pick a latitude and longitude out of a hat, and make the magnitude a random number between 4.0 and 9.0. My first random prediction turns out to be magnitude 8.8 in the middle of the Sahara. And so on. My real predictions, whether based on some hypothesis or simply fabricated cynically, are all plausible. The random ones are much more likely to be nearly impossible. So I can guarantee that at the end of the year, my "real" predictions will have done much better than random.

I am not inventing this method; it has been applied in the past to try to show the validity of a research program.[12] Beware any earthquake predictions that claim to be better than random; the phrase is like a warning light. The real question is whether the predictions are better than intelligent guesswork.

Even without deliberate subterfuge, assessing earthquake prediction methods is problematic. Suppose an earthquake occurs just outside the predicted date range or is just slightly too small. A critic will say that the prediction failed. But the proponent will say that it was so close that by discarding it, something useful could be missed. Maybe if the data had been better interpreted, the prediction could have been more accurate. One can sympathize with the proponent, but once you start allowing near misses to count, how near is near? The scope for dispute is huge.[13]

SPARKS FLY

Perhaps the epitome of the precursor-driven earthquake prediction method is the VAN system, which first appeared in the 1980s. It split the seismological community: some saw a promising new avenue of research; others saw only smoke and mirrors.

The acronym VAN comes from the initials of the three members of the Greek team that devised the system: Varotsos, Alexopoulos, and Nomicos. Varotsos and Alexopoulos were a husband-and-wife team; Nomicos's part was largely in providing computing and mathematical support.

Varotsos was very much the driving force, a role he played well. With a shock of dark hair streaming back from his head and a firm, determined manner, Varotsos looked every bit the scientist-inventor. He would stride into a conference hall trailing television crews in his wake. Varotsos is surely the only seismologist who has ever been featured, or ever will, as the hero in a comic book (in Japan). He also has the honor of being almost certainly the only seismologist to figure personally in an editorial cartoon in a leading newspaper, something I don't think Charles Richter ever managed. But more on that later.

The VAN system was started up in Greece in the early 1980s, with the first field tests in 1981 and a full network of eighteen stations completed in 1983. The driving idea was that transient electrical currents are emitted from rocks once they are stressed beyond a certain degree; because this critical point is below the point at which the rocks will fracture, the current is emitted before the fault actually breaks. These currents were given the name *seismic electric signals,* or SES for short. Each detecting station was made up of a series of electrodes plugged into the ground, constantly monitoring the electrical field in the rocks. Odd-looking disturbances would be interpreted as signals.

According to Varotsos, electrodes couldn't go just anywhere. Certain spots seemed to be sensitive and others not. He also claimed that a site would receive signals from only a certain distance away. So the basis of making a prediction would be as follows: Record an SES and locate the epicenter by determining the area to which the recording stations were sensitive. The size of the SES would give magnitude, and the character of the signal (for instance, was it an isolated burst, or more gradual) would give the lead time from the signal to the earthquake.

Over the rest of the 1980s the VAN team claimed a number of successful predictions, and in the following decade Japanese seismologists took an interest in the method, leading to further experiments in Japan. (And hence the comic book adventures, in a manga comic called *Shonen Sunday*. The two-part series filled eighty pages and was prepared in consultation with Varotsos himself.)

So what could be wrong? From the beginning, relations between the VAN team and other Greek seismologists were strained. While it could have been construed as rivalry over the funding of the project, which at one point was absorbing 40 percent of Greek government spending on seismology, that old problem arose: prediction is difficult, especially about the future. Instead of lodging each prediction with the National Observatory in Athens for validation after the event, VAN recorded its predictions by sending telegrams from one team member to another. Then, after an earthquake occurred, the telegram would be produced, showing the date stamp that proved it was made in advance of the earthquake. The question that was whispered in the corridors at conference sessions was this: How many telegrams were quietly burned when the prediction failed?

This lack of openness also extended to the predictions themselves. Other seismologists found it difficult to see how the VAN team got from a squiggly line on a record produced by one of their stations—which might just be noise from human activity—to a working prediction. And if the system was so obscure that no two seismologists looking at the same SES would come up with the same prediction, it was possible to argue that in fact the success record claimed by VAN was just a series of lucky guesses. A lot of statistical effort went into proving that the VAN record was no better than educated guesswork, and the argument went back and forth for years.

Matters came to a head in 1995, when a special two-day symposium on VAN was arranged jointly by the International Council of Scientific Unions and the Royal Society; it was held in London at the Royal Society's headquarters. Supporters and opponents presented their views, but no definitive answer was found. Observers concluded that while the SES prediction method was anything but straightforward to apply and the statistical evaluation was problematic, it was not clear that there might not be something real underlying it all.[14]

But if the result of the symposium was inconclusive, nature was about to play a hand. In September 1999 a moderate earthquake (5.9) struck on the outskirts of Athens. Normally an earthquake this size in Greece would not attract a great deal of attention, but Athens is situated on a relatively stable block and is mostly earthquake-free. An earthquake large enough to cause damage to the capital caused a great deal of alarm, not to mention killing 143 and leaving some fifty thousand people homeless.

According to VAN, they picked up clear signs of the impending quake. But their failure to actually issue a prediction caused many to wonder what was the good of spending so much money on a system that gives you a prediction only after the event. This was when the newspaper cartoon appeared. It showed a clerk in a lottery office saying, "Yes, Professor Varotsos, your numbers are all correct. But you didn't register your ticket before the draw!"

As if to try to regain the initiative, Varotsos now issued a prediction that a larger earthquake was on its way and would strike central Greece shortly. This caused widespread public alarm, especially since the human impact of the Athens quake had made the Greek population more than usually wary of earthquakes. The scientific community was more skeptical. A committee was convened to evaluate this new prediction. Varotsos refused to attend, and after a few hours of discussion, the committee unanimously pronounced the prediction to be baseless. And of course, no earthquake occurred.

The events of 1999 largely sank VAN in Greece as a credible system, but the team continues to operate out of the Institute of Solid State Physics of Athens University. If anything successful comes out of VAN in the long run, it will probably come from Japan, away from the personalities and controversies that have marked the Greek saga.

WHAT WAS SEEN AT PARKFIELD

The persistent failure of earthquake prediction schemes to deliver the goods has led some seismologists to propose that earthquake prediction is actually an impossible task. There are some grounds for supposing that this might be the case.

It was once assumed that a fault break in an earthquake was fairly straightforward: it starts, it rips along the fault line, it stops. Once it became possible to analyze the process in more detail, seismologists discovered that this is not the story at all. It turns out that all great earthquakes start out as small earthquakes. What typically happens is that a small earthquake occurs on a fault—say, magnitude 2—and this then triggers a larger one a fraction of a second later, which sets off one larger still, and then the rupture escalates into a big earthquake. Why this happens is still a mystery. There doesn't seem to be any reason why some small earthquakes turn into big ones and others don't. Or, to put it another way, how do earthquakes know when to stop getting bigger? This is currently one of the biggest unanswered questions in seismology.

This matters a lot to earthquake prediction. Take the VAN system as an example, which relies on supposed SES emissions before a big earthquake. But if a big earthquake actually starts as a small earthquake, how can this be? There's nothing to distinguish a small earthquake that grows from a small earthquake that stays small.

If all this sounds pessimistic, it's because most seismologists these days are pessimistic about earthquake prediction. The tunnel has been long, and there's still no glimmer of light at the end of it.

What did all those instruments arranged around Parkfield record before the 2004 quake? Precisely nothing. No electrical signal, no radon, no magnetic effect, not a flicker. The inevitable conclusion is that there is simply no precursor that infallibly occurs before an earthquake.

It seems that the best one can possibly achieve is a system that only works sometimes.

It should also be asked if prediction is really all that useful from a social point of view. As with the case of Peru, predicting a major earthquake significantly in the future discourages investment and can have a highly deleterious economic effect. If the prediction then proves wrong, the economic pain has been suffered for nothing. In the regimented communist China of the Cultural Revolution period, it was possible to order mass evacuations at will, but evacuations for false alarms would cause discontent in most countries today.

There's actually a slight irony that so many supposedly successful predictions were issued only after the earthquake had occurred, because many seismologists now believe this is the way forward—to issue a prediction that is guaranteed to be successful precisely because the earthquake has already occurred. To make this useful, there is a little trick involved. And that will be the subject of the next chapter.

9

TWENTY-FIVE SECONDS FOR BUCHAREST

ONE AFTERNOON I RECEIVED A PHONE CALL FROM A JOUR-
nalist in Japan, asking me about an earthquake that had just occurred.
However, when I checked, there was nothing showing on the seismographic records in Edinburgh—the reason being that the closest seismic waves, spreading across the globe, were still, at closest, somewhere
under Siberia. The telephone signal, in contrast, arrived from Japan
almost instantaneously. Similarly, when a strong (but fortunately not
very damaging) earthquake occurred in Virginia in 2011, it was said
that many people in the eastern United States heard about it through
Twitter just before they felt the shaking themselves. One wag remarked
that in many cases, people were busy re-tweeting the news when the
shock waves actually reached them.

It's sometimes hard to appreciate just how fast information travels
today. When a major earthquake occurs, the news often spreads faster
than the actual shock waves.

It was not always so—when the great Lisbon earthquake occurred, news traveled across Europe at the speed of a dispatch rider.
For an earthquake in the Americas, it could be months before tidings of

destruction in Peru could reach the coffeehouses of London and Paris. News of an earthquake in Japan might not arrive at all.

The telegraph was the first major innovation to speed up communication. Electrical telegraph systems were operational by the 1830s, and with the expansion of the telegraph system across the globe through subocean cables starting in the 1860s, news really began to travel quickly.

The telegraph system allowed news of the San Francisco disaster in 1906 to appear in European morning papers the next day. But this hardly compares to the situation today, where nearly everyone carries a phone, many of which are hooked up to the Internet.

It is this sheer speed of modern communication that is starting to take the uncertainty out of the earthquake prediction business.

EARTHQUAKES V. TWITTER

The ideal earthquake prediction system would be infallible, never giving false alerts and never missing an event. Then every bet would be a safe bet—a dead cert. A system that gives 100 percent accurate warnings of an earthquake about to occur. In fact such a system is possible, and even in operation in some parts of the world. It's possible because this system gives a warning only if the earthquake has already happened.

What makes this system different from the earthquake predictors in Chapter 8, who only announced their predictions after the earthquake, is the timing. A day after an earthquake occurs is a bit too late for a prediction to be of much use. However, a window of time exists between the occurrence of an earthquake and when the shock waves actually reach the nearest city. It's small, but not too small to be useful.

It comes down, once again, to the issue of relative speed: the speed of information versus the speed of earthquake waves.

Technically this is not earthquake prediction. *Prediction*, in the sense seismologists use the word, means making a statement about when a fault will rupture. The transmission of information after a fault has ruptured but in advance of the arrival of strong shaking is called *early warning*.

The most striking natural warning is provided by tsunamis. As I have discussed, tsunamis give a clear warning to those who can recognize the significance of the withdrawal of the sea, should they see it. But there is another, earlier signal, as I also mentioned in Chapter 6. If you live on the west coast of Sumatra, feeling protracted earthquake shaking, even if it is not very strong, is a warning. If the shaking lasts a long time, it means the earthquake was large, and if it didn't seem strong, that just means it was distant—out to sea. So a tsunami could already be on its way and due to arrive in half an hour. The inhabitants of the towns along Japan's northeast coast were well warned that a tsunami was coming after the March 2011 earthquake. The unfortunate thing was that many did not make good use of the information.

The relatively slow speed of a tsunami means that a warning can have quite a long lead time. Thirty minutes, or, if you are unlucky, fifteen, is still enough to give people a good chance of getting to a safe place. And that is for a nearby event. When the earthquake is on the other side of the ocean, people will not feel it and will not get the same natural warning from the earthquake shaking. But there is even more time to get a warning out by human means.

The Hawaii Volcano Observatory was started up early in the twentieth century to monitor Kilauea, one of the most active volcanoes in the world. One can tell only so much about what is going on inside a volcano by standing on the edge of the crater and peering in, and the only way of seeing right to the heart of it is through seismology. The movement of magma deep within a volcano's subterranean plumbing generates many small earthquakes, and recording these provides a window into the inner workings of the system. As a simple example: if records indicate that over a period of days or weeks, minor earthquakes that were deep are getting shallower and shallower, it's a strong sign that something is moving upward and an eruption may be imminent.

The observatory was therefore well equipped with seismometers to record small local earthquakes deep in the roots of Kilauea. Inevitably, they recorded anything else that was taking place, particularly large earthquakes elsewhere in the world.

One such earthquake occurred on March 2, 1933. A retired seaman who worked for the observatory, Captain Robert Woods, was about to change the paper on one of the instruments that was installed in the cellar of his house, when the needle started to flicker as a marked seismic signal arrived. Woods watched to see what would happen next. It was almost ten minutes before the next prominent twitch of the needle arrived. So the earthquake was quite far away, he reasoned. It was also big.[1]

He conferred by telephone with one of the observatory's seismologists, Austin Jones. Jones had also seen the waves arriving on an instrument on the other side of Honolulu. They compared notes. The most likely location—the east coast of Honshu, Japan. And since that is a subduction zone, a great earthquake, as this undoubtedly was, meant a very high chance of a tsunami. Bad news for the Japanese, who were probably already inundated by it, but it was also potentially dangerous for Hawaii. They made some calculations and concluded that the wave would arrive in about eight hours. Plenty of time to make preparations.

The harbormaster at Hilo was duly informed, and a message was sent to the other islands that a tsunami was expected mid-afternoon, and people should get ready to be well away from the shore when it arrived. By midday radio bulletins from Japan were already confirming Woods and Jones's judgment; Iwate prefecture had been hit.

The tsunami reached Hawaii just as forecast, and no lives were lost.

WATCHING THE PACIFIC

In 1946 the story was rather different. This time the earthquake, a massive 8.6 magnitude, struck not populous Japan but an obscure spot in the Aleutian island chain, a remote location in Alaska. No one in

Hawaii chanced to notice the seismograms that might have provided warning, and four and a half hours after the earthquake, the tsunami crashed onto the waterfront at Hilo. In the downtown area, people fled in terror as the roaring wave swept up the streets, smashing everything in its path. Many did not make it to safety; 159 were killed.

The earthquakes of 1933 and 1946 provided a clear contrast and a message: tsunami warnings could be given—but only if someone could be relied upon to give them. It was no good taking the chance that someone like Captain Woods would happen to be looking at a recording drum at the right time. Constant watch had to be maintained.

Organizing the task of tsunami watching in Hawaii, day in, day out, required the federal government to take overall responsibility. Since the US government–owned geomagnetic observatory in Honolulu already had a suitable piece of land in Ewa Beach, this became the home for the new Tsunami Warning Center, which was up and running by 1949.

Developments continued to be driven by events. Only after the tsunami from the great 1960 earthquake in Chile wreaked damage as far away as Japan was it realized that making tsunami warning a matter of international cooperation would be a good idea. The Tsunami Warning Center became the Pacific Tsunami Warning Center, with the responsibility to provide alerts of probable (and definite) tsunamis to all the nations around the Pacific Rim, a task it continues to the present day. Whereas at one time warnings were sent out by telex to designated recipients, in the age of the Internet, anyone can visit the PTWC website (http://ptwc.weather.gov/) and read the latest alerts.

Since the 2004 Indian Ocean tsunami, the PWTC has expanded its operations to cover the Indian Ocean as well, and initiatives are being developed to provide warnings in the Atlantic and Mediterranean.

While this system works well for tsunamis, earthquakes are a much more difficult problem. Tsunamis are relatively slow and damaging over long distances. Earthquake waves are fast and do most damage at short distances. So while a warning system might be possible, the system cannot depend on seismologists poring over records to decide exactly where and how large an earthquake is. Even then, the amount of warning is going to be very limited and measured in seconds.

FASTER THAN AN EARTHQUAKE

Surprisingly, the idea of an earthquake warning system is not entirely new; it was actually proposed as long ago as 1868 by a correspondent for the *San Francisco Evening Daily Bulletin*. At that time, the significance of the San Andreas Fault was not realized, but a series of earthquakes had been occurring near Hollister, 120 kilometers away. Why not install a seismic sensor at Hollister and connect it to a telegraph? The telegraph at the San Francisco end could be connected to a huge bell atop city hall that would automatically ring as soon as an earthquake was detected, giving warning to the citizens. The idea had no takers in 1868, and it would be more than a hundred years before anyone realized that an early warning system for earthquakes could actually work.

The best place to build an early warning system is one where the danger from earthquakes is pretty much confined to a single source, and one that is some distance from the main population center. A classic case is that of Romania.

The seismicity of Romania is uniquely odd. Most of the country experiences only low or moderate levels of earthquake activity. The exception is the Vrancea region in the northeastern part of the country. Here, in a very restricted area, large earthquakes, sometimes exceeding 7 in magnitude, have occurred regularly throughout history. This small but active source has another peculiarity—it is deep. The big earthquakes repeatedly occur 100 and 150 kilometers down.

Usually it's a good thing when earthquakes are deep, because it means that even if you are standing at the epicenter, you are still a long way from where the energy is being released, and its force will be diminished somewhat by the time it reaches the surface. However, the dense mantle is more efficient at transmitting energy than the crust. When an earthquake occurs at shallow depth, the waves must travel entirely through the crust to reach the nearest city. With a deep earthquake the waves travel diagonally through the mantle.

As a result, deep earthquakes can be felt at great distances. Vrancea earthquakes are typically felt in Moscow—in fact, it is sometimes assumed that any earthquake felt in Moscow must have originated in Vrancea, 1400 kilometers away. The city at real risk, however, is the capital of Romania—Bucharest. Although it is one hundred kilometers to the southwest, Bucharest is close enough to have been badly damaged by the larger Vrancea earthquakes, most recently in 1970.

This is an ideal situation for early warning, because we know exactly where the threat is. So, to install a system, seismologists need only a strong motion sensor somewhere in Vrancea, a radio transmitter connected to it, and a receiver system in Bucharest. When an earthquake happens, the sensor detects strong shaking and triggers the radio transmitter, which then sends a signal. The signal is received in Bucharest almost immediately, and the shock waves arrive in Bucharest about twenty-five seconds later.[2]

So—twenty-five seconds' warning. How useful is that?

Clearly you can't evacuate a city in twenty-five seconds. Arguably, this is not entirely a bad thing. Suppose you had a day's warning of a major earthquake. The chaos and panic and traffic jams that would ensue as everyone attempted to flee might actually worsen the impact of the earthquake when it struck.

Short timescales favor automatic systems. A computer can accomplish a lot in twenty-five seconds, including the safe shutdown of vital systems like power stations, closing off gas mains, activating safety systems in factories, and so on.

A classic example is the safety system used by the Japanese railways. Tokyo is linked to most of the other major cities in Japan by high-speed rail lines, the *shinkansen* (literally, new trunk line). The name is also used for the trains that speed along the tracks as fast as three hundred kilometers per hour. The operator, Japan Railways, takes pride in the safety record of the system. Since the first high-speed trains started operation in the 1960s, not a single life has been lost through collisions or derailment. But in a country so earthquake-prone as Japan, maintaining this record is a challenge.

Japan began to develop an early warning system after the Kobe earthquake in 1995, although it took twelve years and around half a billion dollars before it was operational. A huge number of seismometers are linked in together in a network run by the Japanese Meteorological Agency (JMA). Warning signals are generated automatically when the network picks up a major earthquake. Since the larger Japanese earthquakes, like that of March 2011, occur offshore, the system usually can provide at least twenty to thirty seconds of warning, and sometimes more. For the rail system this is more than enough to trigger an automatic braking system to slow the trains down or bring them to a standstill if need be.

Which is what happened after the 2011 earthquake. A friend of mine was traveling from Kyoto to Tokyo at the time. He remarked afterward that while he was used to British trains stopping unexpectedly in the middle of nowhere, he didn't expect it in Japan. He soon realized the reason.

And it was just as well the trains were stopped. Aside from the destruction wreaked by the tsunami, in many places bridges were cracked and unsafe, or ground slopes failed in a series of minor landslips, tearing and mangling rail lines.

A train full of passengers running onto a destroyed section of track at hundreds of kilometers an hour is horrible to contemplate. Altogether, the track was unsafe or destroyed at about twelve hundred places up and down the shinkansen network; any of these could have derailed a train with disastrous consequences. The investment in warning systems paid off.

THE CITY ON THE LAKE

Mexico City, like Bucharest, is another special case. Despite the nearby volcanoes, this part of Mexico is not particularly seismic. The big problem lies three hundred kilometers to the south, off the coast of the state of Guerrero. Here is another subduction zone—the Cocos Plate, a relatively small, almost triangular, plate, is pushing north-northeastward and diving to destruction under the southernmost part of the North American Plate.

On September 19, 1985, a great earthquake—8.1—struck along the coastline. I remember it rather clearly, because it happened in the early afternoon in Edinburgh. In those days, before computer screens were universal, there was a tall chart recorder in a two-meter-high cabinet in a corridor at my institute. This displayed the readout from three seismometers radio-linked from hilltop sites in central Scotland on a continuous paper loop.

These particular instruments were designed to detect high-frequency waves from nearby earthquakes, and they were not all that sensitive to the long rolling surface waves that come in from a large earthquake a long way away. But that afternoon the pens were tracing big sweeping waves across the width of the paper. Soon a little group of people was standing around the cabinet watching. Clearly this was bad news for someone, and a few quick measurements showed where.

It was particularly bad news for the inhabitants of Mexico City because of factors partly geological and partly historical. When the Spanish conquistadors arrived in Mexico under Hernán Cortés, they found that the Aztec capital, Tenochtitlan, was cleverly built on a lake, using a system of causeways connecting islets in Lake Texcoco, which made for an excellent defensive arrangement. After the conquest Tenochtitlan remained the capital and eventually became the modern Mexico City.

But the city could not remain distributed around a few small islands—so eventually the lake was filled in and built over. A geological cross section under the modern city would show a basin filled with soft lake sediments.

When the seismic waves arrived, it was like shaking a bowl of jelly. Worse, waves that entered the basin got trapped and started bouncing around within it, combining with fresh waves coming in and making a complex pattern of strong vibrations. When a building sways in an earthquake, it acts like a huge inverted pendulum. The taller the building, the longer the swing it makes; this is known as the natural period of the building. If the wavelength of the seismic waves is the same as, or close to, a building's natural period, the shaking is greatly amplified. This phenomenon is known as *resonance*. To see it in action, take a plastic ruler and hold it vertically between your finger and thumb at the bottom. Move your hand back and forward. If you make very small movements, the ruler won't sway much. If you make a large movement, it won't sway much either. There will be just one size of movement where the ruler starts to really sway a lot. That's the resonant frequency. In the case of Mexico City, buildings between about six and fifteen stories high turned out to be particularly sensitive. Low-rise houses and very tall buildings were not particularly affected, but those buildings in that critical height range (and there were a lot of them) suffered severely. The exact death toll is uncertain, but it numbered in the tens of thousands.

Like Bucharest, Mexico City is in an ideal situation for earthquake early warning. There is one major source of hazard—the coastal subduction zone—and it's a long way away. The peculiar geological vulnerability of the city, sitting on a pile of soft lake sediments, means that faraway earthquakes are still hazardous.

Soon after the earthquake of 1985, just such a system was installed.[3] Twelve detector stations were placed along a three-hundred-kilometer stretch of the Mexican coast, all linked by radio to a central control unit in Mexico City. To try and prevent false alarms, the system was set up so that two sensors at different locations had to record significant shaking before an alert was triggered.

Because of the large distances involved, residents of Mexico City would be able to receive warnings of as much as a minute or even more, so it was decided at the outset that public alerts would be issued, rather than restricting the warning system to automatic shutdowns.

The planners decided to use radio messages to issue alerts instead of sirens or bells. Many special radio receivers were distributed to emergency centers, public offices, schools, and radio stations. The last were configured so that in the event of an alert, regular radio programs would be automatically interrupted with a special sound and the message "Alerta sismica! Alerta sismica!"

The system had its first (and so far only) real test on September 14, 1995. The earthquake, 7.3 in magnitude, was far below the scale of the 1985 event, but it was large enough to count as "major" and cause an alarm. In fact, although it caused considerable damage in some coastal towns, there was no damage in Mexico City, though the shaking was strong enough to be frightening.

The earthquake occurred just after eight in the morning. Many had just arrived for work; others were still commuting. Some schools had already begun classes. Across the city the normal sounds of music from radio sets were cut off by the repeated announcement of "Alerta sismica!" Seventy-two seconds later the shaking started.

Experiences on that day were mixed. In some places, especially the schools, which were well prepared, evacuations were orderly and no one panicked. Some residents carefully turned off the gas and made their way calmly to prearranged assembly locations. The metro system responded automatically: trains slowed and stopped as soon as they arrived at the next station, though no information was relayed to passengers. But it was not clear how many Mexico City residents had heard the signal. Only about 10 percent of the population was likely to be listening to the radio at that time in the morning. And even if those who heard the warning passed the message on to others nearby, that is still a fraction of the population.

The Mexican experience has shown up a number of problems. The best estimate of the number warned in 1995 is a little less than half the population. If a warning given when everyone was up and awake reached only that number, even fewer would catch the warning for an earthquake in the middle of the night. While there has been talk of special radios that turn themselves on when an alert is broadcast, getting these distributed in sufficient numbers would not be easy. New technologies may provide the answer; the Mexican government is now trying an extension of the system to provide warnings using BlackBerries.

An alert that no one hears is one problem, but it's also not much use if people do not know what to do when they do receive it. This is a general problem with warning systems: scientists like to tackle the problems of designing the technology, but effectiveness requires that care also be taken in communicating to the public what the warnings mean. It's all very well to detect a tsunami-producing earthquake and compute its probable travel path, but making that knowledge useful may depend on something as simple as a person on a bicycle with a whistle to get people away from the shore.

This was realized at the outset in the design of the Mexican alert system, and after it was set up, the authorities distributed many leaflets around the city, and special broadcasts advised people how to respond when they heard an alert. But the leaflets were issued only once, and the information broadcasts were short-lived. Unless important information about warning systems is repeated frequently, it will be forgotten. It's easy to let your guard down when other troubles seem more pressing.

Further complicating matters, the Mexico City system has not been completely reliable. A magnitude 6.7 quake in 1993 should have sounded an alarm but didn't, and there have been at least two false alarms. These were upsetting, but no one was actually hurt as a result, and no material damage occurred.

WARNING BY PHONE

The system developed in Japan to protect the shinkansen also offers warnings to the general public. These alerts are transmitted by television and radio. Radios that are in sleep mode, rather than actually turned off, can be awakened by a special tone in order to deliver the warning. Any computer system that is always on can be configured to receive warnings through its Internet connection.

New developments in phone technology offer another way forward. An application has been developed for the latest generation of smart phones in Japan that will pick up messages from the early warning system. This will likely be much more effective than self-activating radios, but it is still not (at least, not yet) going to give universal access. Even then it has drawbacks, as constantly accessing the alert system is likely to drain phone batteries rapidly. But it demonstrates that new technology can alter things in ways not thought of a decade ago.

From the testimony gathered, it seems that millions of Japanese did receive as much as a minute's warning of the impending shaking on March 11, 2011. What is hard to quantify is how many lives were saved as a result of such alerts. The farther people were from the offshore rupture zone, the longer the warning period they got—but the farther away they were, the weaker the shaking was and the less they needed warning. The death toll from the Tohoku earthquake and tsunami combined is estimated at more than fifteen thousand, but only a small percentage of those resulted from the earthquake—most died in the tsunami. The best estimate of the number killed by the earthquake itself is about 230. Given that this was a magnitude 9 earthquake, the damage caused by earthquake shaking seems to have been moderate, a tribute to Japanese engineers and builders.

The future usefulness of the warning system will vary according to circumstances. In the case of an onshore quake like the Kobe earthquake, which struck directly under a major city, the warning is likely to arrive at the same time as the shaking, making it of little benefit to the public. But even in such cases, stopping trains that are traveling into the disaster zone will still save many lives.

Earthquake warning systems generated a lot of interest in the seismological community about ten years ago, which led to some being installed in what might be considered surprising places, such as Istanbul, which is only a short distance north of the North Anatolian Fault, the great strike-slip fault that shears almost the entire length of northern Turkey and poses the greatest threat to the capital. Under the most favorable circumstances, the Istanbul system might generate only eight seconds of warning.

JUAN DE FUCA AND SAN ANDREAS

Early warning systems are not much of an advantage in the intraplate conditions of the eastern United States. The eastern seaboard is not immune to seismic activity, as the 2011 Virginia quake demonstrated. Larger and more damaging quakes have occurred in the past, notably the 1886 Charleston earthquake, which may have had a magnitude as high as 7. What no one knows is whether there is something special about Charleston that suggests that any future event will occur in the same place, or whether the next equivalent earthquake could occur anywhere. If anyone does get advance warning of a future strong earthquake in the eastern United States, most likely, as in 2011, it will come by means of an instant messaging service like Twitter.

The US city that probably stands to benefit most from earthquake early warning is Seattle. The Pacific Northwest has similarities to Japan. The main threat to Seattle is from a great subduction earthquake offshore. The culprit here is the Juan de Fuca Plate, which is moving eastward to destruction under the North American Plate. It's actually a rather small plate; it probably used to be a lot larger, but most of it has already gone to destruction deep in the mantle, and what we see today is the small remnant.

Small it may be, but it's still large enough to produce magnitude 9 earthquakes, the last one being on January 26, 1700. While we have no contemporary eyewitness accounts of that earthquake, we do have some of the tsunami, which swept across the Pacific and hit Japan. By analyzing the Japanese descriptions of the tsunami and collecting evidence of the impact on nature in what is now Washington State—from the tree ring record, for example—seismologists have been able to reconstruct the earthquake. And evidence of earthquake-triggered underwater landslides off the coast suggests that the 1700 quake was not a one-off event. In fact, one can trace about twenty major earthquakes in the last ten thousand years. The pattern of these is interesting: they seem to occur in clusters. A quake will occur about every three hundred years within each cluster, but the gap between clusters is much longer. With the last quake just more than three hundred years ago, it could be that the next one could be soon—or maybe the 1700 earthquake was the last in that cluster. And remember Parkfield: earthquakes don't run like clockwork.

The consequences of a repeat of 1700 for Seattle would be severe. The Alaskan Way viaduct is the main north-south artery through the city. As its name suggests, it is an elevated roadway, and it is a good candidate for the title of scariest roadway in North America, since it is all too obvious that any major earthquake is guaranteed to bring the crumbling edifice crashing down. Just walking under it is unnerving. It was proposed in 2008 that the viaduct be closed and replaced in 2012, but for a while it seemed unclear whether this would actually happen. If the politicians didn't agree to demolish it, the risk was that an earthquake might do the job for them. Work is underway, with completion expected in 2013.[4]

An earthquake early warning system in Seattle could provide as much as a few minutes' notice before the heavy shaking arrived. It might not be possible to clear the whole viaduct of rush-hour traffic, but it would certainly be a help. Seismologists in the Pacific Northwest have their eye on the prospect. Early in 2011, John Vidale, director of the Pacific Northwest Seismic Network at the University of Washington, commented that for about $40 million such a system could be up and running in five years or so.[5] All that is needed is the political will to make it happen—preferably before the next earthquake, rather than after it.

But mention earthquakes in the United States, and everyone's mind turns first to California. And here, early warning systems are currently being developed.

On the face of it, California is not the most promising location for early warning. Cities have grown up right on top of some of the most dangerous faults. While the system proposed for San Francisco in 1868 might have been able to warn of earthquakes originating in the Hollister area, it would have been of little use in 1906, when the fault break was much closer to home.

All the same, warning could be advantageous in some cases. Los Angeles, for instance, suffers from two principal hazards: damaging but relatively moderate quakes directly beneath it (Long Beach 1933 and Northridge 1994, to name two examples), and much larger quakes on the San Andreas Fault or a related system farther away. The danger from the distant faults arises partly because, like Mexico City, Los Angeles lies in a basin that seismic waves can bounce around in. Modeling has shown that in the worst combination of circumstances, a great earthquake starting at the southernmost point of the San Andreas and breaking northward could result in protracted, strong shaking right across LA. But in such a case a warning of as much as a minute might be possible.

Once again, the safety of the transport system is spurring local leaders to action. In Northern California the operators of the Bay Area

Rapid Transit (BART) are particularly concerned. A major quake during rush hour, with sixty packed trains in service, would be a nightmare.

But so far progress is slow. There is a system, which the project coordinator memorably described in late 2011 as "matured to the point where it's sort of duct tape and bailing wire." That is, it's operational but relatively crude and certainly not at the stage where it could be hooked into a public warning system using television and radio as in Japan. The project, currently a cooperative effort of Cal Tech, the University of California, Berkeley, and the USGS, has so far consumed about $10 million, but much more is needed for a statewide system: perhaps $80 million to install and $20 million a year to actually operate.[6] It is not clear where that money is going to come from.

Early warning systems are no panacea. What they are good at doing, like putting the brakes on the bullet trains, they do well. But they work best when the hazard is from well-known sources that are a good distance away. Bucharest. Tokyo. Seattle.

When the fault is beneath your feet, they are no help at all. In that case you need protection to survive, not warning.

10

EARTHQUAKES DON'T KILL PEOPLE, BUILDINGS DO

THE EIGHTEENTH-CENTURY HISTORIAN EDWARD GIBBON IS famous chiefly for his monumental *History of the Decline and Fall of the Roman Empire,* which appeared in six volumes between 1776 and 1788. His stock still stands high among professional historians today, and students are sometimes warned, should they believe they have an original idea, to check that Gibbon did not think of it first.

I have never heard Gibbon cited for his thoughts on earthquake-safe construction, but this is an omission that needs to be set right. In his characteristically florid style he explains:

> In these disasters, the architect becomes the enemy of mankind. The hut of a savage, or the tent of an Arab, may be thrown down without injury to the inhabitant; and the Peruvians had reason to deride the folly of their Spanish conquerors, who with so much cost and labour erected their own sepulchres. The rich marbles of a patrician are dashed on his own head: a whole people is buried under the ruins of public and private edifices, and the conflagration is kindled and propagated by the innumerable fires which are necessary for the subsistence and manufactures of a great city.[1]

A more succinct way of putting it is a slogan often repeated by seis-mologists: earthquakes don't kill people—buildings do.

This is only a slight exaggeration. It may come as a surprise to those whose idea of an earthquake is the B-movie cliché of a crevasse full of lava, but earthquake disasters are as much man-made as natural. To reduce the loss of life in earthquakes, it's necessary to have a clear idea about why earthquakes kill people in the first place. It's not by shak-ing them up. That can be fatal if the shock induces a heart attack, and perhaps a few dozen are killed this way each year. Thousands more are killed by shaking buildings that come down on people's heads.

The chances of surviving an earthquake depend very much on whether the building (if one is indoors at all) is strong enough to sur-vive without collapsing and, to a lesser extent, if parts do fail, how much injury their failure will inflict on those inside.

The danger is really from collapse, either complete or partial. It doesn't matter so much if the walls of a house are deeply cracked, so long as the roof doesn't fall in. And if the roof does fall in, what it is made of makes a difference. This is precisely what Gibbon was com-menting on. Tents and grass huts are safe in an earthquake region. But when Spanish colonists imported Spanish building styles into the Andes, unaware that the country was far more earthquake prone than Spain, disaster was inevitable. Gibbon was probably writing with the 1746 Lima earthquake in mind, the deadliest earthquake during the colonial period in Peru, with more than five thousand fatalities.

In the case of any earthquake with a high death toll, three factors have usually converged: the area affected was densely populated, the predominant building style was heavy and weak, and the earthquake occurred at a time when most people were indoors—usually nighttime. This last factor is often overlooked. The time of day can make a huge difference in the statistics of survival. In a traditional village society, if a strong earthquake occurs during the day, everyone is out in the fields. The houses in the village collapse, but they are mostly empty, so casualties are few (recall Sefidabeh in Chapter 1). Exactly the same earthquake occurring at night will be a different story. The houses will be full and the inhabitants will be killed as they sleep.

It goes without saying, though, that cities cannot be built out of tents or grass huts. We need ways to build houses that are safe when shaken by a strong earthquake.

"Resistant to earthquakes" does not mean "earthquake-proof." There's no such thing as a completely earthquake-proof building that will never be damaged. So it is important to understand the first principle of earthquake engineering. One can't stop a building from being damaged, but one can try to prevent it from collapsing. The key objective is to prevent deaths. So if a building fails in ways that don't threaten the occupants, that can be considered a success. If a wall splits in two, it's alarming for the occupants, but it won't kill them. If a wall collapses, it will. It's good if engineering measures can reduce or prevent damage in the next earthquake, but to some extent that is a bonus. The main thing is to make sure people don't get killed.

Engineers sometimes quote from the Code of Hammurabi, the ancient Sumerian king and lawmaker, who ruled that if a house fell down and killed the occupant, the builder should be killed in turn as punishment (and as a warning to other builders). While no one is suggesting that modern builders should meet such a fate, the king's ruling reinforces the idea that preservation of life is the engineer's goal in earthquake-prone regions.

In a way, prediction and protection are two sides of the same coin. In the first case one says, "So long as buildings are empty when the earthquake strikes, it doesn't matter if they fall down; people will be safe." The second approach goes, "So long as buildings don't fall down when the earthquake strikes, it doesn't matter if they are not empty; people will be safe." While both prediction and protection have the primary goals of reducing loss of life, protection can also reduce economic losses, and prediction does nothing for that.

Also, while the track record of prediction in reducing earthquake casualties is rather feeble and perhaps limited to one case of any significance (Haicheng), the success of good earthquake engineering practice is well established. Journalists are notorious for not liking to report good news. When an earthquake kills tens of thousands of people, the event gets massive coverage. When tens of thousands are not even hurt, because the buildings they lived in were well designed, well built, and well maintained and rode out the earthquake safely, that gets little coverage.

One example is the earthquake of September 25, 2003, which occurred just off the coast of Hokkaido, in northern Japan. While it wasn't in the same league as the March 2011 quake, at 8.3 it was still a great earthquake and caused extensive damage to the transportation network as sloping ground gave way in a series of slumps and slides, tearing up roads and rail lines. While it was duly reported in the Western media, it wasn't lead headline news, for one simple reason. The death toll was exactly one—reportedly a man hit by a car that was flung across the street.

CARD HOUSES

To understand how to build safe houses, it's necessary to know why buildings fail in an earthquake in the first place.

The first rule of building construction: it must stand up under its own weight. The first force a building has to withstand is the downward force of gravity. If you have ever tried building houses from playing cards, you'll know that this is not always easy.

An object is stable so long as its center of gravity (the point representing the average position of its mass) is directly over some part of the area of its base. This is why it's easier to knock over a tall thin object than a short fat one. A slight tilting of the tall object will displace its high center of gravity from its position over the area of its base, whereas a short fat object will take a lot of tilting before sufficient displacement occurs. It's also why a wrestler crouches down while watching his opponent; a crouching position is more stable because one's center of gravity is lower. In contrast, someone carrying a heavy weight on her shoulders is more likely to overbalance and fall because her center of gravity is higher.

Experimenting with building blocks demonstrates this principle. The simplest houselike structure you can make is two blocks of wood balanced on end for the walls and a crosswise block balanced on the top of them for a roof. This will stay up indefinitely under vertical forces only, that is, under gravity. It may not be so stable if you give it a prod from the side. This imparts a new force, a sideways, or lateral, force. Can the building cope with this?

Quite possibly not. The lateral force causes the wall to tilt, its center of gravity is no longer supported, and the wall overbalances. The roof is now unsupported and falls in. Down comes the structure.

This is not so different from real life. Any building or structure has to cope, first and foremost, with the vertical force of gravity. Earthquakes add a new dimension. When the ground shakes, a sideways force acts on the building. If the building has no resistance against lateral forces, it will fail. The art of protecting buildings against earthquakes is, to a large degree, making them resistant to sideways forces.

It's possible to learn quite a lot about this basic principle by playing around with models made out of household items. Building toy houses out of old-fashioned wooden building blocks is a sort of slab construction. This does have real-world parallels.

There was a common type of construction in the Soviet Union known as precast concrete panel. The basic idea was that you could manufacture concrete walls in bulk offsite, truck them in as needed, and assemble apartment blocks quickly and inexpensively. The problem was that this was rather like building card houses, except that the walls were concrete slabs. There is a similar type of construction in the West called tilt-up construction; it is generally used for any of the large shedlike retail outlets in suburban malls. The prebuilt walls are put flat on the ground and then tilted up into their vertical position, hence the name. The stability of these structures depends on how well the walls are secured when tied in.

Huge estates of these Soviet-era precast buildings were erected. Around Moscow and Leningrad, two of the least seismic cities in the world, this didn't cause any real problems. But this was not a type of construction that fared well when subjected to earthquake shaking, and when reinforcement was lacking or inadequate, the buildings could became death traps as the heavy concrete slabs caved in on the occupants. In 1988 an earthquake struck the city of Spitak in Armenia, where many apartment blocks were built of this precast type, rather unsuitably, given that Armenia is a seismically active area. The buildings crumbled. The death toll of thirty thousand was appalling, but not surprising in the circumstances.

Most large modern buildings, though, are based around a frame made of girders. A simple way to experiment in frame construction is with drinking straws. Take four and fasten the ends together with modeling clay or something similar so that you have a square. Push one side of it, and you will find it deforms easily into a lozenge shape, with virtually no resistance. Do the same thing by making three straws into a triangle, and you will get quite a different result. The triangle is impossible to change out of shape. This is a rather interesting fact: some shapes are stronger than others. In particular, rectangles are weak, triangles are strong.

The same thing applies when working in three dimensions. With some more straws the square can become one side of a cube. Push this about a bit, and it's just as weak as the square was—you can push it all sorts of ways. Make a three-dimensional triangle—a tetrahedron—and, again, you can't bend it unless you actually break the straws, which takes much more pressure.

Now imagine that the cube is a building, the straws are girders, and the sides are filled in with brick walls. If the framework is distorted, turning from a square into a lozenge, clearly the brick walls are going to burst apart, and the building will not survive.

Unfortunately, it suits human beings a lot better to live in square houses than in triangular ones. So a basic problem is that the shape most convenient for living in also happens to be the worst in terms of surviving earthquakes.

SEND REINFORCEMENTS

A range of techniques is available to protect buildings against earthquakes, and obviously the most straightforward is reinforcement. Putting diagonal bracing elements in a steel frame building is a simple way to give it lateral strength. It's a way to turn those weak squares into stronger triangles without having to build tetrahedral buildings.

In modern buildings where the frame is exposed, this diagonal pattern can often be seen, and not just in areas where earthquakes are expected. Wind also damages buildings by applying lateral forces, so reinforcement intended to make a building resistant to strong winds helps protect against earthquakes, and vice versa. Although the United Kingdom is not a very seismic country, when occasionally modest earthquakes do occur, they cause noticeably little damage. Part of the reason is that anything built to withstand stormy winter gales sweeping in from the Atlantic is going to have some innate resistance to earthquakes as well.

Putting in diagonal crossbeams is only one type of reinforcement, and while it is well suited for steel-frame buildings, it is not so useful for some other types of construction, such as masonry.

Another useful technique is what is known as ring-beam reinforcement. A building is like an inverted pendulum—the base is fixed to the ground, while the top can sway about. Hence, as Flamsteed noticed, earthquake shaking is more strongly felt on upper floors of buildings. Damage commonly starts at the top of a building for more or less the same reason. Therefore, it's the top of a building that you need to hold together.

The ring beam is typically made out of reinforced concrete and runs right around a building (commonly a house or a small office) just below the roof line. It's a bit like holding a bunch of flowers together with an elastic band—put the band around them at the top of the stems, and you'll certainly stop any stalks from drooping. An advantage of this method is that it is neither difficult nor expensive, and it can be applied retrospectively to existing buildings (this is called retrofitting) fairly easily and without it being too unsightly in the case of historic buildings.

An even less expensive alternative to the ring beam is to run a steel cable around the top of the walls. It's still essentially a ring beam, just made of cable. I have heard engineers argue that the usefulness of the cable ring beam is overrated, but tests show that it is an effective way of strengthening existing structures that were built without earthquakes in mind—one recent application has been in reinforcing bridges on major interstate highways in the US Midwest.

Flexibility in a building can also be advantageous. For domestic construction it's hard to beat wood-frame houses for earthquake resistance. Their flexibility allows them to ride out earthquake shaking and return to their initial position. Rigid additions, like a brick chimney stack, may break, but losing a chimney stack is not a calamity if the building itself doesn't fall in on its inhabitants.

This is why the 1988 Spitak earthquake and the 1989 Loma Prieta (California) earthquake make such a contrasting pair—they occurred a year apart and were of about the same magnitude. Whereas Spitak killed thirty thousand, the death toll from Loma Prieta was sixty-three. The low-density wood-frame houses of California could not be more unlike the precast concrete apartment blocks of Soviet Armenia. Wood-frame houses can still be destroyed in an earthquake, and some were in the Loma Prieta quake. But destruction in this case means the whole house was wrenched off its foundations. Photos of such damage are interesting to look at. It is clear enough that the house is wrecked. There's no way to repair it; just tear it down and rebuild. Whereas a destroyed building in Armenia often looked like a pile of rubble, a destroyed Californian home still looked pretty much like a house. It's easy to imagine that an occupant of such a house would be thrown terrified to the floor, but aside from some bruises, they would be in good shape once the earthquake was over.

The same is true of traditional housing types in other parts of the world. People often note that the Japanese built houses out of wood precisely because it gave them earthquake resistance. A lot of traditional housing in sub-Saharan Africa is virtually indestructible in earthquakes, so easily does it sway about and return to its proper shape. A common style of traditional African housing consists of small circular huts made of poles held together with mud. This will be topped with a light thatched roof of palm leaves. This is the style, for instance, in the Machaze district of Mozambique, which was hit in 2006 by an earthquake of magnitude 7.0—the same size as the Haiti quake four years later. Most Mozambique houses rode out the shock, and the total death toll was just four, with twenty-seven injured.

So it would be easy to say that using wood is the way to build safe houses, but this is facile. People have always used whatever building materials come to hand most easily, and in many parts of the world this is not wood. It would be pointless to recommend the adoption of wooden-house construction in arid regions that have no trees and where the little wood available is needed for fuel.

Other factors besides building material can make a structure more or less susceptible to earthquake damage. Shape is an important feature. Like squares and rectangles, different shapes affect how a building behaves. To give a simple example, take an L-shaped building, oriented roughly north-south and east-west. The wing that runs north-south is going to be stiffer in that direction than east-west, because it is easier for the building to sway sideways than along its length. The other wing is going to be stiffer in exactly the opposite direction for the same reason. So whichever direction the earthquake shaking comes from, one wing is going to vibrate more strongly than the other. This is inevitably going to cause trouble where the two parts of the building join, and probably the building will literally come apart at the seams. In contrast, the huts that resisted the Machaze earthquake so well were aided by their symmetrical circular shape.

The more irregular a building is, the more complex will be the interaction of its different parts vibrating in different ways, and the more likely it is to be damaged. This also applies to the internal structure. A common design for an office building is to have a rectangular shape with open-plan spaces on each floor and a stairwell in one corner. The problem here is that most of the building is relatively flexible compared to the stairwell, which is stiff. The building is pinned by that stairwell, so in an earthquake it twists around and breaks up. If the stairwell had been in the center, the building would have been much safer.

So as a general rule, the more symmetrical and regular the building, the less it will be damaged in an earthquake. The extreme case is the building that is a perfect cube. This is the ideal shape for earthquake safety, but it makes for rather boring buildings. The different factors have to be balanced. While choice of building material has to be tailored to what is available, shape has to be a compromise between the requirements of earthquake safety and ordinary functionality and aesthetics. So when the architect comes up with an imaginative design featuring all sorts of strange overhangs, it is the engineer's job to inject a sense of reality and devise some arrangement that accommodates both what the architect would like and what is reasonable from the point of view of structural safety.

One of the most famous earthquake-resistant buildings in the world is the Transamerica Pyramid in downtown San Francisco. Although its distinctive tapered pyramidal shape arose primarily from considerations of light and shade, it is also good for earthquake resistance, since it is both symmetrical and bottom heavy. The building was equipped with instruments that gauge its response to earthquakes, and during the 1989 Loma Prieta earthquake it suffered no damage. The whole building swayed smoothly, with no twisting. Ideal performance.

A common problem in modern buildings is the infamous "soft story." This is when a lower floor is substantially weaker than the rest of the building, usually because it is mostly open space. There can be various reasons for this. A common one is car parking. If parking space has to be provided, the usual place for this is at ground level—what is called "tuck-under" parking. The parking area will be as open as possible, with as few walls as possible, to maximize access.

This means, though, that this whole floor of the building has little strength, with all the other heavier, stiffer floors balanced on top. Thinking in terms of models, it's a bit like balancing a book on four matchsticks. It can be done, but the result is not going to be very stable.

Come the earthquake, the inadequate supports give way, and the whole story simply vanishes as the upper floors descend on it en bloc. Anything or anyone in the soft story at the time is crushed. The result can be quite strange to see. The building may look as if it has not been badly damaged. Some rubble lies around the base, but there's not much sign of cracking or fallen walls. Then someone tells you, "That was a seven-story building." You count the floors and find only six. The upper floors are sitting on a thin layer of rubble where the ground-floor level once was.

Apartment buildings with tuck-under parking were hit particularly hard in the Northridge earthquake of 1994; most had been built in the 1960s and early 1970s before building code provisions strongly discouraged the practice. More than 2,700 apartment complexes were damaged in the quake, and two hundred with tuck-under parking collapsed, killing thirty-three people.

Parking is one reason why soft stories get built; another is to have a nice open atrium at street level. Hotels can be offenders here. At street level the hotel owner would like a big, open reception area with glass walls and lots of marble, as well as some nice big rooms for holding conferences and wedding receptions. All the bedrooms are on the upper floors, which are braced with numerous internal walls. The ground level becomes a soft story and is crushed to nothing in an earthquake as the upper floors crunch down on it.

The extreme case is the building on stilts—nothing but empty space at ground level and the whole building supported on columns, often a misguided mixture of modern aesthetics and the desire for more parking space. Usually the columns are far too slender (again, for aesthetic reasons) to support the upper floors under lateral forces, and the whole thing comes crashing down.

Sadly there is no way to prevent designers or builders from making the same mistakes over and over again. Designers seem to always want to put a huge, weak, open atrium on the ground floor because it looks good. Or builders use substandard materials to save money—such as whoever built the office buildings that collapsed in Taipei and were found, after an earthquake in 1999, to have domestic rubbish stuffed into the walls as infilling.

Avoiding building features that will obviously cause problems in an earthquake is one way of making a building safer. Ironically, though, another tool in the engineer's kit is to install deliberate weaknesses. The idea here is to control how a building gets damaged, and this relies on the principle that the weakest link is always first to fail. The engineer includes zones in the building's design that are designed to fail without jeopardizing the safety of the whole structure. In the process of failing they will soak up the energy of the earthquake. Therefore the building will not collapse, and life will be preserved. It's a neat idea, all the more so for its paradoxical principle of deliberately designing parts of the building to be weak.

ARTIFICIAL EARTHQUAKES

Theory is all very well, but engineers need a way of actually testing ideas for designs without waiting for an earthquake. A certain amount can be done with computer modeling, but ultimately, the best test is to build a design and then shake it and see what happens. In other words, it's necessary to simulate an earthquake.

This is done using something called a shake table—basically, a platform with some sort of motor underneath that can make the whole thing shake. The idea of testing buildings in this way is not a new one—as early as the end of the nineteenth century the Japanese were testing columns by shaking them. And the Byzantines, way back in the sixth century, may possibly have used some sort of shaking table in the design of the final version of the Hagia Sofia basilica in what was then Constantinople.[2] Certainly, the survival of that building to the present day in one of the more earthquake-prone cities of the world is a testament to the care of the architects who designed it.

Shake tables come in various sizes and designs. Obviously, for testing a full-size building, the shake table has to be pretty large itself. But smaller ones can be used to test models of structures. The smallest can sit on a desk. A colleague of mine built a tiny one that is essentially a loudspeaker lying on its back with the diaphragm uppermost, connected to a plate. It's just large enough to put a couple of toy buildings on top—plastic model buildings with the walls held together with plastic putty rather than glue. Earthquake recordings converted to audible frequencies can be played through the loudspeaker, which causes the plate to vibrate, and if the shaking is loud enough and long enough, the building falls apart. Such a toy has no engineering use, but as an educational tool for demonstrating earthquake shaking it has proved popular with visitors when the institute opens its doors to the public once a year.

Serious shake tables are powered by an arrangement of pistons that can move the surface of the table in all six directions: back and forth,

side to side, and up and down. With computer control they can reproduce any recorded earthquake motion exactly. Being able to reproduce
past earthquakes is extremely useful. The engineer can ask questions
like, "What would happen if my building had needed to withstand the
2011 Tohoku earthquake? Would it have stayed standing?"—and then
find the answer by taking an accelerometer recording of the strong motion produced by that earthquake, relaying it through the table, and
having the pistons produce the exact sequence of bucking and rolling
that actually occurred. Or one can modify the recording. "Suppose the
shaking from the 2011 Tohoku earthquake had been 20 percent stronger?" Boost the motion and run the test again. It's certainly a better way
of testing than building the structure for real and hoping it survives
when the next earthquake comes along. This way, design features can
be evaluated in a safe testing environment.

Most of the largest shake tables are in Japan. The biggest one is
in Miki City, east of Osaka. It is 20 × 15 meters in size and can handle
buildings weighing as much as twelve hundred metric tons. The largest US shake table, in San Diego, measures 12.2 × 7.6 meters but can
take heavier loads—as much as two thousand metric tons. The largest
European table is 6 x 6 meters and is located in Saclay, on the southwestern outskirts of Paris.

A few of the smaller shake tables are built as museum exhibits, so
that the public can get a taste of what a large earthquake feels like un-
derfoot without the danger and inconvenience of actually being in one.
One such is in what used to be the Geological Museum in London, now
the Earth Sciences wing of the Natural History Museum. When first
installed, it reproduced a Venezuelan earthquake from the 1960s. The
interesting thing about this earthquake was that when it occurred, a local schoolteacher had been giving her class singing lessons, which she
was recording on a tape recorder. When the earthquake struck, she and
her class ran out in a panic, leaving the tape recorder running, which
captured the rumbling and crashing sounds of the earthquake. Today,

with security cameras and modern cell phones everywhere, such recordings are commonplace, but back in the 1970s they were rare, and being able to demonstrate not just what an earthquake feels like but also what it sounds like by playing the audio recording in sync with the ground motion, was something special.

Some years ago I visited one of the larger shake tables in Italy, just outside Rome, on an afternoon when tests were being run to explore how useful steel cables were in protecting historic buildings like churches, of which Italy has rather a lot. A simple stone arch in traditional masonry had been reconstructed on the shake table, and a single cable had been run around it near the top. Various earthquakes were being replayed on the shake table, and a number of sensors placed around the stones were recording the result.

Finally, all the tests were done. "OK," said the guy in charge, "all done for today. Let's just knock it down now."

So the technician ran the strongest earthquake he had through the machine to knock down the arch and save having to dismantle it. By now the arch had already taken a bit of punishment, and the masonry blocks were mostly just resting on top of each other, the mortar having already been broken. Under the final strong shaking, the whole arch bucked and swayed about. The stone blocks snapped apart and together again like a movie director's clapper board. Anyone standing under that archway would have felt sure that her last moment had come, that in mere seconds she would be buried in stone as the archway came down in pieces.

But when the shaking stopped, the arch was still standing. Even when it seemed impossible that it could survive, the cable held it together. It was an impressive demonstration.

TOPPLING COLUMNS

A select group of earthquake engineers keep a bag perpetually packed. They always need to be ready to catch the next plane to who-knows-where. These are the members of the national earthquake field reconnaissance teams. Their task is to head to the scene of any earthquake disaster as soon as possible to conduct the post-mortem, as it were. And they have to do this as soon as possible, because they need to examine buildings before the rubble is cleared away. They can't start at once, because search-and-rescue missions must have priority. But as soon as the rescue work is finished, the seismologists and engineers move in to investigate, to find out not only what went wrong with the buildings that collapsed but also what went right.

In the United States one major organizer of such missions is the Earthquake Engineering Research Institute, a national nonprofit technical society (see www.eeri.org/). The United Kingdom has its own team, the Earthquake Engineering Field Investigation Team, a joint venture of industry and academia that has been carrying out missions all over the world for decades (see www.eefit.org.uk). Other countries, including France, Germany, and Italy, also have their own teams. Always, the aim is to learn from experience. Discovering why buildings failed in one earthquake can help ensure that techniques are improved before the next one.

Given the large number of buildings that were either built before earthquake design codes were introduced or that used out-of-date and inadequate codes, engineers on such missions are not only interested in how new buildings performed. The effectiveness of retrofitting measures is also important.

One example is to be found in the history of freeways. While these are the arteries of urban development in modern cities, once they reach a certain density, they become messy tangles of concrete spaghetti—criss-crossing overpasses and underpasses. The extensive networks of bridges and overhead routes can lead to vulnerable structures that are stiff and top-heavy. Freeways have collapsed in a number of modern earthquakes; one of the more spectacular examples was the collapse of the Hanshin Expressway in the 1995 Kobe earthquake. This overhead route was balanced on a single central row of concrete columns that toppled over, sending a long stretch of the roadway crashing down. Such failures have the potential to result in a high loss of life, depending on the time of day and state of the traffic at the time.

One earthquake that features strongly in the annals of earthquake engineering was the San Fernando earthquake of February 9, 1971, also known as the Sylmar earthquake. Its impact on the transportation network was dramatic: twelve overpasses collapsed onto the roads below. Because the earthquake occurred at 6 A.M. local time, only two people were killed in these collapses. For the local population it was a lucky escape. The engineers had more to think about. For one thing, the earthquake was remarkable for the extremely strong ground motion it produced—the highest (at more than 1g) that had ever been captured on instruments at that time (see Chapter 5) . But also it happened during a period when the earthquake engineering community had been rapidly developing ideas.

One challenge faced by the engineers in rebuilding after San Fernando was what to do about collapsing freeways. The main problem appeared to be the columns—they were too weak. That was not just an issue for the rebuilding of the damaged sections; other parts of the freeway system might still be at risk in the next earthquake. The simple solution seemed to be jackets—strengthening pillars by putting concrete casings around those that might be at risk.

Once the immediate task of rebuilding was done, engineers began a concerted program of reinforcing bridge columns around the Los Angeles freeway system. The test of the result came sooner than expected. Twenty-three years later, the Northridge earthquake struck only about eleven kilometers southwest of San Fernando, and it hit many of the same roads and bridges. For the reconnaissance teams a principal task was to find out to what extent the freeway system had been improved.

The initial signs were not too good: media coverage gave wide exposure to pictures of collapsed bridges and tales that, with the road system out of action, Angelenos had at last discovered the train. This was misleading; although some bridges did come down, by and large the road infrastructure came through the Northridge earthquake well. The worst damage was on Highway 118, closest to the epicenter and therefore the section most exposed to shaking.

The teams found that the performance of the bridges varied, but those that fared least well turned out to have identifiable design flaws—where Highway 118 crossed Mission Boulevard and Gothic Avenue, and Bull Creek Canyon Channel, for instance, the road was laid out in such a way that its mass was irregularly distributed. When shaken, the whole roadway twisted about, breaking away from the supporting columns. Once again the problem was shape. Interstate 10, said to be the busiest road in the United States, was also badly damaged in places. Ironically, one section of I-10 that collapsed (at La Cienaga Boulevard) had been scheduled to have the reinforced jackets fitted in a few months' time.

In bridge after bridge that was examined, the jacketed columns showed no sign of damage. But was this too good to be true? Perhaps the columns were damaged under the jacket and the engineers just couldn't see it. It was important to find out. CALTRANS, the agency responsible for maintaining the system, selected some pillars where the earthquake shaking had been strongest and removed the jackets. Underneath the engineers found no damage. The scheme had been a success.[3]

BOUNCING BUILDINGS

While reinforcement is an obvious way to make buildings resist earthquakes, earthquake engineers have some less obvious tools in their repertoire.

Since buildings are set in the ground, if the ground moves, the building is damaged. But if you could somehow insulate the building from what the ground does, you could remove most, if not all, of the earthquake's capacity to damage the structure. This is called base isolation. The principle is similar to the suspension system in a car—springs partially absorb the jolts as the car travels over a rough surface. The technical term is *decoupling*—in a decoupled system, parts move independently, even when joined.

Putting a building on springs would be difficult, but there are three alternatives: putting it on rollers, putting it on a sliding plate, or putting it on rubber pads. In all cases this means redesigning the foundation. In a conventional building the whole structure is tied into the foundation. In a base-isolated building this is no longer the case. Imagine that the bottom of a foundation pit is lined with giant rollers like so many enormous logs, and then the building is constructed so it sits on top of the rollers. The sliding plate and rubber pad versions work similarly. Very early designs suggested the use of ball bearings or even talcum powder in foundations to prevent the ground movement from being communicated to the building. Of these alternatives, rubber pads are the most commonly used and work rather like giant shock absorbers (although technically, they deflect the shock rather than absorb it).

The idea was tentatively advanced early in the twentieth century but was largely dismissed as unworkable until the late 1970s, when it was realized that the great advantage of using rubber bearings is that they are inexpensive and easy to make, don't involve moving parts, and last a long time.

The technique did not catch on at once but has become increasingly popular, especially in Japan. It's not suitable for all building types; it's more appropriate for midrise construction than high rise, and it remains too expensive for use in residential housing. But it's still a useful tool for the earthquake engineer.

Perhaps the most bizarre idea for decoupling housing from ground movement was the *takht-i push* method of construction experimented with in Iran in the 1960s. The idea here was to do away with foundations altogether and build a perfectly spherical house. If the ground shakes, the house can just roll around without sustaining any damage. However, I imagine the experience of the occupants would be anything but pleasant. At least two of these were built in northern Tehran (and perhaps only two), but no earthquake hit the city in this period, so they were not tested. Definitely not an idea that one would want to export to the hilly streets of San Francisco. Even on the flat, it would be worrying for the occupants if the people on the top floor decided to throw a party.[4]

Another strange, but rather more practical, idea is what's known as the *tuned mass damper*. The objective here is to change the way a building vibrates in an earthquake by suspending a huge mass in the middle of it. This damps the vibration, hence the name. It's a bit like the way a mute makes a violin quieter. It's also useful for reducing vibration from wind motion and even the vibration from feet marching across a bridge. The most spectacular example is probably in the Taipei 101 building—this is one of the tallest buildings in the world, and it's in one of the most active earthquake zones, so it needs protection.

The suspended damper is gigantic—a 730-ton ball suspended high up inside the structure, occupying four entire floors. It was impossible to lift such a heavy weight when the building was being constructed, so it had to be assembled in place from forty-one huge steel plates. It has yet to be tested in a major Taiwanese earthquake, but shock waves from the 2008 Wenchuan (Szechuan) earthquake sent the ball swinging, much to the consternation of visitors lining the observation gallery.

THE COST OF SAFETY

Although there are examples of earthquakes where many people died because poorly engineered structures failed, the fact remains that whenever the death toll from an earthquake is exceptionally high, it's most often the case that the majority of deaths occurred in villages where the structures were erected by local builders using materials to hand and following practices handed down from generation to generation.

Reducing the number of deaths from this type of construction in developing countries is a totally different problem from protecting modern buildings in California or Japan. Expensive solutions are of no use in poor backwoods villages. And papers in engineering journals are of no use to a village house builder who can't even read. What's needed here are extensive programs that send educators into the villages to engage with the builders and show them ways of making the houses they build safer. A friend of mine has seen builders in a seismically active part of India laying bricks for house walls by placing the bricks upright on their sides instead of on their faces. The bricks go further that way, and the builder saves a few rupees. But the house becomes a death trap in the next strong earthquake. The builder probably never even thought about that—which is why education is so important. No one is going to recommend building steel-frame, base-isolated houses in remote parts of India. Solutions have to be appropriate to the local situation, in terms of cost and cultural sensitivity.

This is a major issue in Nepal, one of the poorest and least developed countries, which also sits in a major earthquake zone along the southern margin of the Himalayas. The Udaipur earthquake in 1988, on the Nepal-India border, was a wake-up call. Despite a fairly moderate magnitude of 6.6, the devastation was widespread because of the low building standards. More than a thousand people died, mostly in Nepal.

This triggered a series of initiatives, in which I had a small part, to try to improve seismic safety. I spent several months in Nepal in the early 1990s examining the situation; speaking to local experts, politicians, and planners; and drawing up guidelines for the work that needed to be done. In 1992 work began on the Nepalese National Building Code Development Project, as it came to be known, and the following year the National Society for Earthquake Technology—Nepal (NSET) was founded as a focus for seismic safety initiatives.

By the late 1990s the effort was in full swing. A national building code had been drafted, complete with updated earthquake hazard maps. Safety leaflets had been published. Schoolteachers were being trained in earthquake safety. And assessments were being made of the safety of the most important structures: hospitals and schools.

By the following decade, the vitally important training programs for masons were in progress throughout the country. There is even a fixture in the country's calendar—January 16 is National Earthquake Safety Day, the anniversary of the worst earthquake to strike Nepal in the twentieth century (in 1934). In 2012 the fourteenth National Earthquake Safety Day was marked by meetings, rallies, and exhibitions. It is heartening to see a country mobilizing to tackle the threat of earthquakes in such a systematic way and gratifying to have personally made a contribution, if only small. NSET's stated goal is to have earthquake-safe communities in Nepal by 2020—a huge task but one that is being confronted with dedication.[5]

Cost, though, is not just an issue in the developing world. Cost is always an issue, especially in major engineering projects. While it is possible to use every technique available to make sure that every building is as earthquake-safe as possible, it's perhaps not good value for money. Would it make sense to duplicate the design of the Transamerica Pyramid, with all its earthquake safety features, in Newark, New Jersey, where the chance of an earthquake is extremely low? Those safety features might just turn out to be a waste of time and money.

Earthquake engineering projects are a partnership between the engineer and the seismologist. It's the engineer's job to decide on the design of a building and which earthquake safety features to include. But what the engineer does not know is when to stop adding safety features or even whether they are needed at all. The building needs to be safe, but how safe is safe in different parts of the world? This is a job for the seismologist.

11

THE PROBABILITY
OF DISASTER

IMMEDIATELY AFTER THE 2010 HAITI EARTHQUAKE, EXTEN-
sive aerial photography was used to assess the scope of the damage in
Port-au-Prince and the outlying towns and villages, a number of which
were inaccessible in the first days after the quake. Some of these aerial
views provided striking images of the destruction. In the central part of
Port-au-Prince the cathedral was clearly visible from its cross-shaped
outline. Then one noticed the shadows inside the cathedral—shadows
cast on the floor of the nave by the window embrasures. The entire roof
was missing.

Perhaps the most emblematic picture of the whole disaster was
the collapsed presidential palace. After an earthquake disaster there
are three absolute essentials: the government must continue to function; emergency services—hospitals and fire stations—must still be in
service; and major communication routes must remain open. The collapse of the presidential palace showed that the Haitian government
had not been able to protect even itself, never mind the rest of the population. The earthquake had effectively decapitated the entire country.
Reportedly what was left of the civic administration convened in the
shade of a tree near the ruined airport in the days after the quake. There

again is the contrast between those two earthquakes in early 2010—
Haiti and Chile—the unprepared country and the prepared country,
as I discussed in Chapter 1. But while the citizens of Haiti might have
been unaware they had an earthquake problem, there were seismologists who certainly knew about it

The Haiti earthquake came as no surprise to seismologists. Since
2003 the tectonics of Hispaniola had been a subject of investigation
for a team of geologists and seismologists led by Eric Calais of Purdue
University. After five years of collecting Global Positioning System
(GPS) data and measuring the strain along the Enriquillo–Plantain
Garden Fault zone, the team concluded that a magnitude 7.2 earthquake was likely if the fault released all its stored energy. They presented
their results at a 2008 conference in the Dominican Republic, but it was
clear that this was not a matter of mere academic interest. A humanitar-
ian disaster loomed. Calais requested, and obtained, private meetings
with representatives of the Haitian government, along with members
of the Haitian Bureau of Mines and Energy. Calais did his best to im-
press upon the officials the seriousness of the situation.[1]

He could not know just how urgent it was. Even if the Haitian government had leaped into action (which it did not, however seriously
officials took the warning), time was running out. A little more than a
year later, the blow fell.

It would be wrong to say that Calais's team predicted the Haiti earthquake or even forecast it. They did not say, and did not know, when the
next major earthquake on the Enriquillo–Plantain Garden Fault zone
would occur, and it was a coincidence and misfortune that it happened
so soon after Calais reported his findings to the Haitians. What the team
did do was identify a danger, something that was likely to happen sometime, and something for which it was necessary to prepare.

We call this seismic hazard (see Chapter 1)—the probability of
earthquake shaking.[2] Assessing hazard means grading places according
to their susceptibility to earthquakes. At a simple level, categories of

high, medium, and low hazard plotted on a map of the United States would show high hazard along the West Coast, grading to medium through the Rockies and low upon reaching the eastern states.

Such simplified maps have their place in national building codes, instructing builders to follow different requirements for safety according to the zone in which the construction site lies. This is usually enough for simple projects like house building, but for complex buildings—skyscrapers, major bridges, nuclear power stations—the engineer needs to know more precisely what sort of hazard he has to design against. It comes back to the question of how safe is safe. Any engineering project has two competing priorities: safety and cost. The skillful engineer can work into his design many earthquake safety features, but these make a building more expensive. Too little antiseismic design and the building is unsafe. Too much and it's wasteful. Clearly a balance has to be struck.

WORST-CASE SCENARIOS

When people first started thinking seriously about this problem in the mid-twentieth century, the answer seemed simple. Find out the worst thing that could happen, design for that, and you are safe. Problem solved. So for any design being prepared, the steps in the analysis would be as follows: First, find the nearest active fault. Second, calculate the largest earthquake that could happen on that fault. Third, imagine that the largest earthquake occurs on the point of the fault nearest the site and work out how strong the shaking will be at that sort of distance. That tells the engineer the maximum force for which he has to design, and nothing worse could ever happen. Or could it?

It turned out not to be as easy as that. There are four big problems.

The first is finding the nearest active fault. This assumes that the geology of any area is perfectly known, which is far from the case even in places where the geology has been pretty well examined. The

Philippines is another region of the world that is strongly seismically active. It forms part of the so-called Pacific Ring of Fire, a roughly continuous belt of earthquake and volcanic activity that runs right round the ocean. Geologists thought they had the active faults of the Philippines mapped out but got a shock when around midday on February 6, 2012, a violent 6.7 earthquake struck the Cebu region near the center of the country.

Which fault was it on? Geologists were embarrassed to admit they had no idea. It didn't match up with any fault on the map.

The problem was that all those faults marked on the geological maps were those that could be seen at the surface. The fault that broke in the Negros Orientales earthquake of February 2012, however, occurred on a fault that did not reach to the surface. It's buried by other rocks. This is known as a *blind fault,* or, in this case specifically, a *blind thrust.* And because it is blind, no one had any idea it was there until it moved.

So finding the nearest active fault is not so easy when the nearest fault is concealed. The same problem came to the fore in New Zealand in 2010–11. The Darfield earthquake west of Christchurch in September 2010 occurred on a completely unknown fault, this time a strike-slip fault, the line of which was covered with gravels and sands washed down from the flanks of the Southern Alps. The city of Christchurch itself fared reasonably well in that earthquake, but devastation struck the following year when a second earthquake, undoubtedly triggered by the Darfield quake, occurred right under the city, causing much more severe damage and many casualties. But no one previously had any idea such a fault existed under the city. Where was that "nearest active fault"? A lot closer than anyone had suspected.

Even if you are sure you know where the nearest fault is, the second problem is calculating the largest earthquake it can produce. It's easy enough to work out the magnitude of the largest earthquake a fault has produced in the past—so long as you have records that go back far enough. But in most parts of the world records have been kept for only a few hundred years, whereas the fault being studied might produce only one earthquake every thousand years. But even if you have good knowledge of the earthquakes that have happened in the past, the possibility remains that a larger one could happen in the future. There is always some doubt.

Geological studies of the Meers Fault in Oklahoma have found traces of large earthquakes from thousands of years ago, but nothing since. In fact the area surrounding the Meers Fault has been practically devoid of earthquakes for as long as the area has been settled. So what is the largest earthquake that the fault could produce tomorrow? Is it still as dangerous as it ever was in geological history, or has it now gone permanently extinct? There are no easy ways to answer such questions.

It gets worse. The third problem is similar to the second one. Suppose you could work out the magnitude of the largest possible earthquake that could occur on the nearest fault—the next task is to determine the maximum strength of shaking at the site of the proposed building. How do we do that?

Well, two things are pretty obvious. You can expect stronger shaking from big earthquakes than from small ones. And the farther away you are, the weaker the ground motion is going to be. Given these two variables, it ought to be possible, with enough data, to work out the expected ground motion for any combination of magnitude and distance. But while you can estimate a value of sorts, in practice the variation is huge. So for an earthquake you think is going to be magnitude 6 and twenty kilometers away, you can work out the average expected ground motion—but your chances of getting the average motion exactly are really quite slim. This is particularly noticeable close to the fault break. Recordings from the Kobe earthquake in 1995 that were taken close to the fault showed variations of a factor of ten between stations only a short distance apart.

So even if you could determine the largest earthquake on the nearest fault, you can't tell exactly what the shaking from it is going to be. All you have is a wide range of potential outcomes.

Last problem. Let's imagine we can solve the other three. We have a site somewhere in eastern North America or northwestern Europe. There might be a fault somewhere close by that might be capable of producing earthquakes. Maybe it could even produce a magnitude 7 quake. We can't be sure, but we can't rule it out. And if an earthquake that big happened, it could conceivably produce very strong shaking. If we are trying to calculate the worst possible case, the maximum, we can end up with an excessively high number.

Well, it could happen. But how likely is it? And is it really worth taking seriously? If we do take it seriously, it costs. One can approach the question from the viewpoint of insurance. In many countries ordinary homeowners' insurance automatically covers a range of perils, but suppose that you have to buy separate coverage for every different peril. Do you buy coverage for fire? Yes, of course, because that's quite common. Houses in every city catch fire every day. What about meteorite impact? Could your house be flattened by a big meteorite impact? Of course it could. Is it likely to happen? No. How many houses have been flattened by meteorites in the last hundred years? So let's save some money and not buy the meteorite policy. (Actually, about sixty houses have been hit by meteorites in the last century, often with damage to the roof, but I'm not aware of any genuine reports of complete destruction.)

This is the concept of acceptable risk. I'm prepared not to worry about meteorites hitting my house because the chance that will happen is so remote I can live with it. I'm not prepared to ignore the risk of fire, because the chance of a domestic fire is much higher.

We make this sort of calculation every day, usually without thinking about it. Crossing the street is a dangerous activity, but we do it anyway, because the risk of getting hit by a car is sufficiently low that we can tolerate it. We don't always make the calculations correctly; some people are nervous about flying, when in fact driving to the airport is far more dangerous. But we make the calculations all the same, even if subconsciously.

So tackling earthquake hazard by always preparing for the worst possible outcome is inherently flawed, because the worst possible outcome may be so remote that the risk is acceptable. A different approach is needed, one that takes into account which risks are acceptable and which are not.

CALCULATING THE ODDS

What we need to do is define what we mean by *acceptable risk*. And that brings us to probability. To say that air travel is less dangerous than driving to the airport is an exercise in probability. Count all the people who have traveled by air in the last decade and all the people who have been in a plane crash, divide the second by the first, and you have the probability of being in a plane crash next time you fly. If the chances of something bad happening, like a meteorite strike, are one in a million, then you may not be too worried. If you play Russian roulette, the odds of shooting yourself are one in six, and that's not very attractive.

Probability is something that many people have trouble really getting their heads around. One famous difficulty is the so-called gambler's fallacy. Imagine a session at the roulette table. The last fourteen balls have all come up red, so the gambler says to himself, "Fifteen reds in a row is such an improbable outcome! I will stake heavily on black." The trouble is that the roulette wheel has no memory. It doesn't know the last fourteen balls have been red; the next spin of the wheel is independent of the previous ones, and the odds for the next spin are still fifty-fifty.

Probability is the solution to all four problems involved in trying to determine a worst-case scenario. It allows a different approach: one where one is no longer concerned with just the worst case. Instead, the seismologist looks at all the possible outcomes. It doesn't matter if you don't know exactly what the fault properties are, or how big an earthquake can be, or how strong the ground shaking will be, so long as you can rank different outcomes by how likely they are. Further, it's possible to calculate just what the likelihood is.

For probability to be useful, it's necessary to go back to the idea of acceptable risk. If one is prepared to define this in terms of an actual number, then it's possible to work out what this represents in terms of an earthquake.

Now, it's easy to take extremes and say that a one-in-a-million chance of something bad is certainly OK, and one in six is not, but it's a bit harder when you have to draw the line precisely. Defining acceptable risk means coming up with a figure where anything more probable is not tolerable. The mere fact that someone is prepared to play Russian roulette means that they will accept a one-in-six chance of shooting themselves. Not many would accept those odds, but assuming someone is desperate enough to try it, would they still try it if the odds were one in five? Or one in four? Or one in three? Somewhere there is a dividing line between where you decide you will play, and where you won't.

When planning for the safety of society, such as writing building codes, a similar decision must be made. Legislators might decide that one in a thousand is OK but one in 999 is not. Who makes the decision depends on the context—and on the consequences of getting it wrong. In building a factory the main consequence of failure is the economic loss to the owner, so the owner makes the call. If the project is a big dam with a major city downstream, in all probability legislation already on the books defines the level of safety required. The decisions about how safe is safe will have been made at some stage by politicians, probably at a national level.

So let's say it's decided that a 95 percent chance that during its lifetime a building will not fail from earthquake shaking is acceptable. The seismologist makes some calculations and concludes it's 95 percent certain that earthquake shaking will not be greater than some value in any fifty-year period (assuming this is the projected economic life of the structure) and passes that value to the engineer, who decides how the building will need to be built to withstand that strength of earthquake shaking.

The seismologist's calculation of earthquake potential brings together many types of information into what is called a *seismic hazard model,* which synthesizes in numerical form anything that could be relevant, including how often, how big, and how strong. The first step is to look at the geological forces, usually the configuration of the tectonic plates that are driving the regional earthquakes in the first place. This is simple if one is analyzing the seismicity of New York, but it is not simple for Haiti, where the crust is broken into a number of small plates that shift in complicated patterns as they are squeezed from both the east and the southwest.

The next step is to look at the major faults—in Haiti, the Enriquillo–Plantain Garden Fault zone is obvious, likewise, the Septentrional Fault, a major structure that runs along the north coast of Hispaniola and was responsible for a huge earthquake that hit the north of Haiti in 1842. But, as with Northridge, it's not always the big famous faults that do the damage, and one has to try to get a handle on what other faults, perhaps even buried ones, might be a hidden threat.

Studies of this kind are no longer carried out only from geological field observation. Increasingly, space-age technology is providing vital information about exactly how rocks are moving—as with the GPS study in Hispaniola conducted by Eric Calais and his colleagues.

If geology provides the why, seismology provides the what—the next key component of the hazard model is the earthquake catalog. David Hume, the Scottish Enlightenment philosopher, once remarked, "All inferences from experience suppose . . . that the future will resemble the past."[3] This holds for seismology. The earthquakes in the historical record are a good indicator of what could happen again, as with the 1770 Haiti earthquake. Even the geological record sometimes leaves clues, such as the records of underwater landslides left by those huge prehistoric earthquakes off the coast of the Pacific Northwest.

Time machines are still in the realm of science fiction, and perhaps they always will be. But suppose for a moment that we had a time machine that could send someone into the past for a little while and bring him safely back. Suppose also that it is hugely expensive to run, and one trip to the past is like sending an astronaut to the moon in terms of cost. Trips would be few, and they would have to be carefully prioritized so that the missions run provided value for money.

How such prioritization would work is an interesting puzzle. There are lots of things that would compete for attention. It would be fascinating to get actual film coverage of Napoleon meeting his final defeat at Waterloo. Or perhaps of the signing of the American Declaration of Independence. One also could settle a few mysteries and see what really did happen to the *Mary Celeste* when its crew vanished without trace in the middle of the Atlantic. These would all be interesting, but would any of them actually be useful in a truly practical sense?

One application that really would have a lasting practical value is in historical seismology. One of the remarkable things about the 1755 Lisbon earthquake is that even after numerous studies no one truly knows which fault produced it. There are at least four candidates and vigorous disagreements in the scientific literature. If it were possible to mount an expedition to the past and send a few seismologists back in time with modern instruments, the issue could be settled. And that would have real benefit in terms of understanding the hazard that faces modern Lisbon.

The worldwide seismic monitoring network has provided copious information about earthquakes, especially since its major expansion in the 1960s, but for slow-moving faults this is not enough. To track down earthquakes of the eighteenth and nineteenth centuries, or even earlier, in the absence of time machines, seismologists need to make use of the skills of the historian.

One of the great characters of historical seismology was the Alsatian geologist and seismologist Jean Vogt. Like many of his compatriots, he had a strong regard for the finer things in life. In his early career the French government loaned him to the Chileans to make a survey of the country's bauxite (aluminum ore) deposits. For some months he spent each day at a bauxite mine and the evening at the nearest winery. As a result he could boast of having literally drunk his way down the length of Chile.

In his retirement he devoted himself largely to discovering information on earthquakes in the historical record. He was the antithesis of the researcher who jealously guards his finds; he happily shared the information he recovered with anyone who needed it. He also had a knack for finding things. He could go into an archive containing thousands of boxes of papers and notebooks, mostly unsorted and without indexes, and unerringly put his hand on the volume that contained copious accounts of past earthquakes.

The past seismicity of the Caribbean was a particular interest of Vogt's. His research notes, hand-scrawled or typed up chaotically on his battered old typewriter, form an invaluable resource for the history of earthquakes of that part of the world. He traveled from island to island, all at his own expense, looking for libraries and archives. The extensive historical records of trade between the Caribbean and New England were one thing that caught his attention. After mulling this over, he set off to archives in Massachusetts, where he triumphantly found hitherto unknown records of Caribbean earthquakes penned by New England traders.

From the record of the past, it is possible to calculate not only where earthquakes have occurred but roughly how often. Beno Gutenberg and Charles Richter discovered in the 1930s (as did Japanese researchers working independently at about the same time) that earthquake catalogs always obey a certain pattern.[4] It's obvious enough that large earthquakes are less frequent than small earthquakes, but the difference is distinctively regular. Basically, for every decrease in magnitude by one unit, earthquakes become ten times more frequent. So for the world catalog of earthquakes, every year on average, for every earthquake of magnitude 8 or greater, there are roughly ten greater than magnitude 7, one hundred greater than magnitude 6, one thousand greater than magnitude 5, and so on.

Wherever you look, the pattern is more or less the same. The multiplier is not always exactly ten but is usually close to it, and it's consistent over the whole magnitude range. This allows seismologists to estimate how likely it is that earthquakes will happen that are larger than ones in the historical record.

So by combining the geological analysis with the average earthquake rate, it's possible to produce a model of where earthquakes occur and how often. The last piece of the puzzle is what effects earthquakes will produce. It would be nice if you could say, "A magnitude 6 earthquake at a distance of twenty kilometers will always produce X amount of shaking," but you can't, and it will never be possible. There are just too many factors. The passage of seismic waves from the breaking fault to a recording station is intensely complex, as waves bounce about and are refracted at every boundary between rock beds. But what you can say is that X will be somewhere in a certain range of values, and some values are more likely than others. And that, as it turns out, is enough. Because the goal is to express hazard in terms of probability, it is sufficient to know the chance that a magnitude 6 quake will cause shaking of X—or any other value.

Once the model has been formulated numerically, it is relatively straightforward to combine the different probabilities—the probability of where and the probabilities of how often, how big, and how strong—at the engineer's site and end up with an overall value that expresses, with 95 percent certainty, the maximum strength of an earthquake at that site during the lifetime of the building.

While this is great for engineers, insurers and planners want a bit more. If hazard is the chance that earthquake shaking will occur, risk is the chance that damage will occur. It is just one more step in the chain. For any particular strength of shaking, one can estimate the probability that buildings of different quality will be damaged severely, lightly, or not at all. Tote that up for an entire portfolio of insured buildings, and the insurer has an idea of what to charge by way of premium to make a reasonable profit and not take too many losses.

Hazard and risk can both be calculated, but as I stressed in Chapter 1, they are not the same. If hazard is only moderate but the vulnerability of structures is high, the risk is high. My Australian friend with the subjective risk scale put the disaster potential for San Francisco at about three out of ten. The place he gave ten to was New York City. Perhaps he was trying to be provocative, but it's not without foundation. New Yorkers certainly don't think of themselves as living in earthquake country—but then collective memory doesn't extend back to 1737. On December 18 of that year an earthquake, probably about magnitude 5, shook the East Coast from Boston to Philadelphia. The surviving data from that time are not enough to fix the epicenter precisely, but the most likely location is on the Hudson River near Manhattan, where several chimneys were thrown down by the shock.

It's hardly necessary to comment on the contrast between New York City as it is today and what it was like in 1737. How many leaning chimney stacks are there today? How many unsecured brick parapets ready to plunge into a crowded street? How would the older bridges fare when shaken by an earthquake for which they weren't designed? The New York City Area Consortium for Earthquake Loss Mitigation has assessed potential losses in New York City from feasible earthquake scenarios at between $85 billion and $200 billion.[5] The total economic cost of Northridge was about $40 billion. Earthquakes in the United States are not a problem only for Californians.

12
STAY SAFE

THE MILLION DEATH EARTHQUAKE MIGHT HAPPEN IN THE near future, and it's only natural to speculate on where. Seismologists look anxiously at Istanbul and Tehran, where a massive death toll would be no surprise in the event of a major earthquake. There is already talk of moving the capital of Iran from Tehran to Qom, a city in a much less hazardous part of Iran. One cannot move the whole city, though.

Of all cities, Istanbul, or Constantinople, as it was formerly known, has one of the longest-recorded seismic histories, thanks to its position as the eastern capital of the Roman Empire. Seven strongly damaging earthquakes are known to have affected the city during the first millennium A.D., a period blank in most earthquake catalogs. One of the strongest historical earthquakes, in 1509, was referred to in the records as the "little Doomsday." Today Istanbul is a city of ten million people. Protecting the city is a challenge. The earthquake engineering community in Turkey is of the highest caliber, but as the collapse of substandard housing during the Izmit earthquake showed, what the local builder does may be what counts.

The case of Istanbul is unnerving for another reason. The North Anatolian Fault, the great strike-slip fault that starts in eastern Turkey and dies out in the middle of the Aegean, has an interesting property. Earthquakes along it tend to occur in sequences, starting in the east and moving progressively west. Each quake, as it occurs, throws more stress on the next section of the fault to the west, which then fails a few years to a decade or so later. It's like a series of dominos toppling. In the historical record this sort of sequence can be seen more than once.[1]

The current sequence began with a 7.8 magnitude event near Erzincan, at the eastern end of the fault line, in 1939. This was followed by quakes progressively farther west in 1942, 1943, 1944, 1957, and 1967. Then, after a lull, the next most westerly stretch of the fault broke in 1999 with the Izmit earthquake. The next stretch of the fault to the west goes straight through the Sea of Marmara, just south of Istanbul. This is the next domino to fall, and it could happen at any time.[2]

Seismologists look even more anxiously at northern India and Nepal. As India is thrust northward by plate movement, scrunching into the Himalayas, strain is building up along the major fault that controls the collision zone. We know from historical records, including Tibetan chronicles only recently uncovered, that a truly great earthquake occurred in this region in 1505, all but wiping out several small kingdoms.[3] This earthquake is not well known. Most earthquake catalogs fail to list it, and when they do, they mention only damage in the north Indian plain, which is distant from the worst-affected region, and fail to convey just how extensively destructive the earthquake was. Compounding these problems is another earthquake that occurred a month previously, near Kabul, and some historians fail to distinguish between the accounts of two earthquakes so close in time. The Tibetan sources record the extent of the destruction in the small mountainous kingdoms that then occupied the southern slopes of the Himalayas: tens of thousands were killed, and around thirty severe aftershocks added to the devastation. One of the local kings was among the casualties, killed as he surveyed building work on his palace.[4]

Now the region is densely populated, and cities like Kathmandu are full of fragile buildings teetering over narrow streets. A report by

GeoHazards International recently described Kathmandu as the most seismically vulnerable city in the world.[5] It suffered most recently during the Nepal earthquake of 1934, which is commemorated each year on its anniversary, January 16.

The 1934 Nepal earthquake has a curious history. At the time Nepal was a closed kingdom; no news entered or left the isolated mountain state. The earthquake damage in Nepal went unnoticed by the rest of the world, but what was noticed was heavy damage in the Bihar region of northern India—which was severe, because the shaking was enhanced by the soft sediments of the Ganges plain. So for many years people assumed that Bihar was where the epicenter was, well to the south of its true location up in the Himalayan mountains. As I discussed in Chapter 10, the Nepalese are making great strides toward earthquake safety, but there is still much to do.

One man who has studied the region extensively, Roger Bilham of the University of Colorado, has divided the mountain front of the Himalayas into ten sections.[6] In the last two centuries, it seems that even the largest earthquakes in the region broke only one section. The 1505 earthquake probably broke two sections and maybe more. The potential consequences of a repeat are staggering. Not just Nepal but much of the densely populated Ganges plain would be strongly shaken. Even worse, a number of major dams now occupy some Himalayan valleys. If a great earthquake were to cause a dam to fail, the resulting flood would be a calamity on its own.

WE'RE ALL IN IT

One would like to hope that the million death earthquake never happens; that unlike in Haiti, measures to reduce casualties are put in place in threatened cities before the next earthquake strikes and are effective. But such preparation will come about only if there is the will to make it happen. Ultimately someone has responsibility for dealing with the threat from earthquakes. In this final section I will discuss where that responsibility lies.

It's not a simple issue; many different people need to be involved. I like to envisage the relationship between the different communities as a large square, such as one might find in a big town or city, with different buildings arranged on the four sides.

- On the north side of the square are the biggest and most impressive buildings. These are the government buildings, and the biggest and grandest is the home of the national government.

National government sets the agenda. If measures are not mandated from the top, they are much less likely to succeed. If the government mandates a program of earthquake safety, it has the power and the authority to make it happen. After all, it is the national government that is responsible for the safety of the realm. It would be surprising if a government decided that national defense was unimportant and all the armed forces could be done away with. It's equally surprising, to my mind, that a government can spend vast sums on military equipment intended to defend the homeland from a threat that will probably never come and next to nothing on defending the population from natural disasters that certainly will come.

One fears that some politicians have a childish delight in acquiring shiny new tanks and missile launchers that make a big show on parade. Spending on seismic safety does not have the same machismo glamour. Then, regrettably, a lot of military spending is prompted less by need than by vigorous lobbying from the defense industry, with deals sometimes corruptly arranged. This siphons off money that could be spent better elsewhere.

National governments need to set priorities. If the Indian Ocean had no tsunami warning system before 2004, that was because the national governments in the region didn't see a reason to commission one. In contrast, the efforts to prepare for earthquakes made at a national level in China are impressive. Some years ago the government established a massive disaster management center in Beijing to coordinate responses to any major earthquake. Immediately after an earthquake occurs, all high-ranking officials with relevant responsibilities assemble in a large room where every seat is equipped with its own communications facilities. From this nerve center all emergency management efforts can be coordinated to maximum effect.

The center got its first major test after the Wenchuan earthquake in 2009. The response to this calamity was impressive and effective, compared to the dithering that one so often sees after earthquake disasters. The army was on the ground for rescue work extremely rapidly, parachuting in to areas made inaccessible by landslides and broken bridges. In managing a disaster a fatal mistake is to waste time developing a plan on the hoof; the plan needs to have been prepared in advance. This is a lesson the Chinese have learned well.

Next to the national government building on the square are the offices of local governments and the various ministries. These are the people who will actually implement the policies decided by the national government. It is here that emergency management plans will be drawn up. It is here that the national building code will be decided on. And it is here that responsibility lies for ensuring the building code is actually followed. Here live the regulators and inspectors who play a vital part in the chain of earthquake preparedness.

To paraphrase Nick Ambraseys, one of the great pioneers of earthquake engineering, in some countries the building code is looked on as a series of instructions; in others, a series of suggestions; and in others, empty words.[7] This is no good. Standards have to be rigorously maintained. And someone has to do the maintaining.

Which brings me to a great problem, one not always appreciated for its negative effect on earthquake safety—corruption.[8] Trying to make progress in a country where corruption is rife is like trying to walk with your legs shackled. Standards can't be maintained if senior ministers are diverting funds to whichever contractor gives them the biggest kickbacks, or if officials can be bribed not to do their job. You can bribe a building inspector—but you can't bribe an earthquake. It's not easy to purge corruption from a political system, but the benefits to society are huge. In some countries today (India, for example), there are signs of mass discontent with endemic corruption, giving hope that maybe a solution can be found.

Although the north side of the square is mostly for government offices, it is also home to nongovernmental organizations (NGOs), which often take over some activities of the state in promoting resilience against natural hazards in developing countries. They also provide relief after an earthquake. After the 2010 Haiti earthquake it was not the dysfunctional government that set up tented villages for the homeless. Here too we find search-and-rescue teams, like International Rescue, who are ready to fly to a disaster zone at a moment's notice and provide expert help in finding and saving those trapped in the rubble.

- The west side of the square has a row of elegant buildings adorned with porticos, cupolas, and pinnacles; on this side of the square live the scientists. These are the people who will provide professional advice to those who need it, so they can make decisions based on the best information. Of course, here are the seismologists who monitor earthquake activity and can say what has happened and what is happening when an earthquake strikes. The hazard specialists advise those working with building codes and frequently wander into the neighboring buildings occupied by the geologists and geophysicists, and, indeed, the mathematicians, statisticians, and even historians—anyone with expertise who can help to quantify where and how often earthquakes occur.

One point worth making is the need for a national, state-funded institute for seismology, especially to manage seismic monitoring. This is a long-term strategic function that depends on continuity and stability. The analysis of earthquake activity depends on having continuous records; it's not something you can stop doing and pick up again a few years later without ill effects. When monitoring is left to a university department or a private company, that stability is harder to guarantee. History offers plenty of examples of an observatory that closed because the guy in charge retired and no one else was interested in keeping it going.

A national institute also provides somewhere for the media to go for authoritative and independent information. Countries without such an institute and no monitoring can be subject to earthquake rumors, scares, and panics in the absence of a national figure who can issue definitive statements to ease unrest.

- The east side of the square is dominated by gleaming modern office blocks that house the construction industry. Here are the earthquake engineers, designing new ways of making buildings safer. Critically, here also are the construction firms who take these designs and make them a reality. The best designs are not of much use if they are badly implemented and shoddily built.

Now we begin to see the interactions that take place across the square. The government offices support the scientists and are in turn advised by them. The scientists advise the earthquake engineers, who collaborate with the builders, who in turn are subject to the oversight of the government inspectors. The end result, we hope, is safer buildings in which to live and work.

And let's not forget the little house around the back of the offices where the humble village builder lives. He may be poor, he may have little education, but he too is vitally important. Most earthquake deaths result from collapsing buildings that were erected by traditional small-time builders. Reaching out to these people is essential.

The tallest towers on the east side of the square house what might be surprising occupants: the insurance and reinsurance industry. The reinsurers are particularly important because, as their name suggests, they insure the losses of the ordinary insurers and are ultimately the ones left holding the financial losses after a disaster occurs. Minimizing damage from earthquakes is very much in the interests of reinsurers, because having to bankroll reconstruction cuts into their profits. But they are not just powerless spectators. Reinsurers can set premiums that are pegged to standards of construction; indeed, it makes sense for them to do so. That gives the construction industry an additional incentive to keep standards high. What they gain from building cheaply, their client stands to lose from higher insurance premiums. The role of reinsurance in promoting earthquake safety is not always widely recognized and perhaps not always fully exploited. It is important, and it's one reason why people from the insurance industry often turn up at seismological meetings.

- Finally, there's the south side. The buildings here are diverse and numerous. This is the public sector. The buildings go back a long way—in fact most of the town is here.

The largest buildings here belong to the media: the radio and television stations and newspaper offices. What happens in the rest of the square is all good in its way, but it is of less utility if few people know about it. The population needs information about earthquakes that are happening, earthquakes that might happen, and measures that are being taken to reduce risk. Information reduces panic. In the aftermath of a strange and scary thing like an earthquake, people are reassured if they see or hear, on television or radio, that there are people who understand what is going on and know what to do. There's no longer any need to rush to the nearest temple to Poseidon and offer up a sacrifice.

Traditional media are not the only sources of information in a disaster today. The advent of new forms of media has saved lives. After a major earthquake those alive but buried in ruins often have difficulty alerting potential rescuers. Assuming a wounded survivor is capable of shouting at all, rubble has a deadening effect on sound; hence the need for listening devices to try and locate those still alive in the wreckage. Being able to communicate by phone certainly improves the odds of being found. Tools such as Twitter and Facebook suddenly take on new importance as a means of communication in the chaos following an earthquake disaster. And not just for the trapped—those who survived the earthquake without injury (but may be homeless) can communicate quickly with friends and family outside the disaster zone. One recent development is the Random Hacks of Kindness initiative, in which software developers are being brought together under the sponsorship of Google, Microsoft, the World Bank, and the National Aeronautics and Space Administration to try to find new and innovative uses of modern technology to reduce the peril of natural disasters.

Also on this side of the square are the schools. Teaching children good practices in seismic safety is advantageous for the entire community. It's not just so that they will grow up to be educated adults—when children go home at the end of the day, they will pass the message on to their parents. This is often a much better way of communicating information to families than by addressing the adult community directly—especially in societies where literacy is low. Parents are likely to pay attention to what their children tell them, especially when it concerns the safety of the entire family. Earthquake safety initiatives tend to put great emphasis on reaching out to schools to teach the basics about earthquakes and how society can be protected from them.

One British family who moved to Japan had a sobering experience. The parents found a school for their two children and met with the head teacher shortly after their arrival in Japan. As the parents were shown around, the head teacher said, "Of course, before your children can start here, we'll need to receive your earthquake plan."

The parents were baffled. "Earthquake plan? What's that?"

"Oh," said the head teacher, "it's quite simple. Suppose a major earthquake occurs and you are both killed. We need to know what to do with your children if that happens. That's the earthquake plan."

Neither parent had really thought about earthquakes in quite such a stark way. The school brought the issue home to them in a real, practical way.

The south side also has many small buildings—whole neighborhoods stretch off into the distance, and that's where you and I live. We're all in this together, and everyone has a role to play in maximizing safety. We need to make sure our houses are as good they should be— by keeping up with their maintenance and making sure, for instance, that we don't allow old chimneys to deteriorate and become hazards.

But many simpler actions can make a big difference if an earthquake does strike—particularly for those living in areas where earthquakes are common and to be expected. I well remember the first night I spent in a hotel in Japan. As I dumped my bag in the room, I spotted a flashlight clipped to the head of the bed. I'd never seen that before in a hotel. "Here," I thought, "is a society that takes earthquakes seriously." It's only a little, inexpensive thing, but having a flashlight at hand, should an earthquake occur in the middle of the night and all the electricity shut off, might make all the difference between finding your way to safety and a dangerous fall.

Look around your home and imagine what would happen if it were shaken violently. Do you have heavy objects stored on high shelves? If they fell down, could they hit someone? Do they need to be up there? It's possible to get little shake-proof clips for fastening down things like computers, televisions, and other heavy electrical items and stopping

them from being tossed across the room in an earthquake. Little preparations can make a big difference.

In my analogy of the town square I described who lives on each
side: government, science, industry, and public. The traffic across the
square is just as important, because the people in all those buildings
have to interact to keep things on track. The politicians need to set
standards for the builders and fund the scientists. The scientists have
to keep everyone supplied with the information they need. Politicians,
builders, and scientists all need to communicate with the media to get
the message across—and so on.

So whose responsibility is it to make society safe from earthquakes?
It's society's responsibility. We're all in it together. And we all need to
know where we stand and what we should do.

Knowledge is power. Knowing what to do can be the difference
between life and death. Instinct may tell you to run outdoors when
a building starts shaking, but that often results in being struck by a
cascade of stones and tiles falling from the outside of the building.
Children in California and elsewhere are taught to "duck and cover" for
good reason: the safest place to get to in an earthquake, assuming you
are indoors, is beneath a good solid piece of furniture.

Knowing what to do means knowing to turn the gas off after an
earthquake. Knowing what to do means knowing the risk of fire. There
may be gas leaking from somewhere. Lighting a cigarette is a bad idea.

Knowing what to do means knowing that if you are outdoors when
an earthquake strikes, it's best to stay outdoors and keep away from
buildings and steep slopes that might become landslides. Aftershocks
might occur at any moment, and a weakened building could collapse
in any fresh tremor. Let the civil authorities decide which buildings are
safe to enter.

Knowing what to do means knowing what it means when the sea
strangely recedes. It means knowing not to go and investigate, but to

clear the beach and get to high ground. Knowing this can be the difference between life and death.

Earthquakes are strange and uncanny things, striking suddenly without warning. One moment you can be doing something utterly normal: loading the washing machine, checking your email, feeding the cat, whatever. The next moment all that stability, all that normality, is gone, and the solid unmoving earth is suddenly moving violently.

The first step toward protection is to know the enemy. Knowledge is power.

Stay safe.

NOTES

CHAPTER 1 SCREAMING CITIES

1. This account is largely based on work by Professor James Jackson and colleagues at Cambridge University. See especially J. Jackson, "Living with Earthquakes: Know Your Faults," 8th Mallet-Milne Lecture, *Journal of Earthquake Engineering* 5, supplement 001 (2001): 5–123.
2. For historical earthquakes in Iran, see N. N. Ambraseys and C. Melville, *A History of Persian Earthquakes* (Cambridge: Cambridge University Press, 1982).
3. V. Marza, "On the Death Toll of the 1999 Izmit (Turkey) Major Earthquake," in *Proceedings of the European Seismological Commision XXIX General Assembly* (Potsdam: 2004).

CHAPTER 2 WHAT *IS* AN EARTHQUAKE, ANYWAY?

1. For a good overview of early theories about earthquakes, see E. Oeser, "Historical Earthquake Theories from Aristotle to Kant," in *Historical Earthquakes in Central Europe,* ed. R. Gutdeutsch, G. Grünthal, and R. M. W. Musson (Vienna: Abhandlungen der Geologischen Bundesanstalt, 1992), 48:11–31.
2. Aristotle, *Meteorology,* trans. E. W. Webster (Adelaide: eBooks@Adelaide, 2007), 2:8.
3. R. Boyle, "A Confirmation of the Former Account, Touching the Late Earthquake Near Oxford, and the Concomitants Thereof," in *Philosophical Transactions of the Royal Society of London* (1666), 1:179–81; J. Wallis, "A Relation Concerning the Late Earthquake Near Oxford, Together with Some Observations of the Sealed Weatherglass, and the Barometer, Both on that Phaenomenon, and in General," in *Philosophical Transactions* (1666), 1:166–71.
4. J. Flamsteed, *A Letter Concerning Earthquakes: Written in the Year 1693* (London: Millar, 1750).
5. For a full account of the Lisbon earthquake and its intellectual aftermath, see T. D. Kendrick, *The Lisbon Earthquake* (London: Methuen, 1956).
6. I. Kant, "Further Observations on the Earthquakes that Have Been Periodically Observed," in *Complete Works* (Berlin: Akademie-Textausgabe, 1968), 469.
7. J. Michell, "Conjectures Concerning the Cause and Observations on the Phenomena of Earthquakes," in *Philosophical Transactions* (1761), 51:566–634.
8. The story is told in R. H. Grapes and G. L. Downes, "Charles Lyell and the Great 1855 Earthquake in New Zealand: First Recognition of Active Fault Tectonics," *Journal of the Geological Society* 167 (2010): 35–47.

9. C. Lyell, *Principles of Geology* (London: Macmillan, 1868), 2:88.

10. Among others, P. L. Fradkin, *The Great Earthquake and Firestorms of 1906: How San Francisco Nearly Destroyed Itself* (Berkeley: University of California Press, 2005), gives a good full account of the disaster.

11. H. F. Reid, *The Mechanics of the Earthquake,* vol. 2 of *The California Earthquake of April 18, 1906, Report of the State Investigation Commission* (Washington, DC: Carnegie Institution of Washington, 1910).

CHAPTER 3 JOURNEY TO THE CENTER OF THE EARTH

1. C. Riedweg, *Pythagoras: His Life, Teaching and Influence,* trans. S. Rendall (New York: Cornell University Press, 2005).

2. C. Davison, *The Founders of Seismology* (Cambridge: Cambridge University Press, 1927), 195.

3. A. L. Herbert-Gustar and P. A. Knott, *John Milne: Father of Modern Seismology* (Tenterden, UK: Norbury, 1980).

4. The experiment was first carried out at Killiney Beach, then Dalkey Island, and subsequently on Anglesey, in Wales. Mallet was an important pioneer and probably the first person to suggest that faulting produced earthquakes and not vice versa, though the crucial text is obscurely worded. An account of his varied career can be found in R. C. Cox, ed., *Robert Mallet, F.R.S., 1810–1881, Papers Presented at a Centenary Seminar 22 Clyde Road, Dublin, 17 September 1981* (Dublin: Institution of Engineers of Ireland, 1982).

5. Thanks to Ina Cecić for sending me one of these stamps. Other seismological stamps I know about bear the likenesses of Emil Wiechert (Germany) and Torahiko Terada (Japan). There may be others.

6. The seismograms today reside in the archives of the British Geological Survey, Edinburgh.

7. Whenever a nonscientist today comes up with a wild and woolly idea that the professionals reject out of hand, the nonscientist may fall back on the "Wegener defense": "You reject my idea because I'm not a geologist, but Wegener wasn't a geologist and he was right." To which the riposte has to be, "You may not be a geologist, but also you are not Wegener."

CHAPTER 4 TRACKING THE UNSEEN

1. R. Feng and Y. Yu, "Zhang Heng's Seismometer and Longxi Earthquake in AD 134," *Acta Seismologica Sinica* 19 (2006): 704–19.

2. A. Bina, "Ragionamento sopra la cagione de' terremoti ed in particolare di quello della Terra di Gualdo di Nocera nell' Umbria seguito l'A. 1751." Costantini e Maurizj (Perugia: 1751).

3. R. M. W. Musson, "Comrie: A Historical Scottish Earthquake Swarm and Its Place in the History of Seismology," *Terra Nova* 5 (1993): 477–80.

4. J. Forbes, "On the Theory and Construction of a Seismometer, or Instrument for Measuring Earthquake Shocks, and Other Concussions," *Transactions of the Royal Society of Edinburgh* 15 (1844): 219–28.

5. Credit for first use of the word seismology in print goes to Robert Mallet.

6. D. Milne, "Notices of Earthquake-Shocks Felt in Great Britain, and Especially in Scotland, with Inferences Suggested by These Notices as to the Causes of the Shocks," *Edinburgh New Philosophical Journal* 31 (1841): 259–309.

7. An earlier invention of a horizontal pendulum instrument by Lorenz Hengler in Munich was not designed for the purpose of detecting earthquakes; see J. Fréchet and L. Rivera, "Horizontal Pendulum Development and the Legacy of Ernst von Rebeur-Paschwitz," *Journal of Seismology* 16 (2012): 315–43.

8. J. Wartnaby, "Seismological Investigations in the 19th Century, with Special Reference to the Work of John Milne and Robert Mallet" (Ph.D. thesis, University College London, 1972), is an excellent resource on the early history of seismometry, sadly unpublished.

CHAPTER 5 HOW BIG? HOW STRONG?

1. The 2009 edition of the *Associated Press Stylebook,* used by nearly every daily newspaper in the United States and most of the weeklies, advises, in its entry for *Richter scale,* "No longer widely used. See *earthquakes.*" The *earthquake* entry takes up the better part of two pages, advises journalists to report the magnitude of the quake, and offers a long explanation, with examples, of what that means. This is only partly correct—the name was never used as a scientific term.

2. J. Milne, *A Catalogue of Destructive Earthquakes, A.D. 7 to A.D. 1899* (London: British Association for the Advancement of Science, 1912).

3. D. Drake, *Natural and Statistical View, or Picture of Cincinnati and the Miami Country, Illustrated by Maps* (Cincinnati: 1815), 233.

4. Ibid., 236.

5. Ibid., 236–37.

6. M. Sarconi, *Istoria dei fenomeni del tremoto avvenuto nelle Calabrie, e nel Valdemone nell'anno 1783* (Naples: Luce dalla Reale Accademia della Scienze, e delle Belle Lettere di Napoli, 1784).

7. P. N. C. Egen, "Über das Erdbeben in den Rhein—und Niederlanden von 23 Feb. 1828." *Annalen der Physik und der physikalischen Chemie* 13 (1828): 153–63.

8. The history of the development of the Rossi-Forel scale can be most conveniently found in C. Davison, "On Scales of Seismic Intensity and on the Construction of Isoseismal Lines," *Bulletin of the Seismological Society of America* 11 (1921): 95–129.

9. R. M. W. Musson, G. Grünthal, and M. Stucchi, "The Comparison of Macroseismic Intensity Scales," *Journal of Seismology* 14 (2010): 413–28.

10. G. Mercalli, "Sulle modificazioni proposte alla scala sismica de Rossi-Forel," *Bollettino della Società Sismologica Italiana* 8 (1902): 184–91.

11. A. Cancani, "Sur l'emploi d'une double echelle sismique des intesitès, empirique et absolue," *Gerlands Beitrage Geophysik* 2 (1904): 281–83.

12. A. Sieberg, "Über die makroseismische Bestimmung der Erdbebenstärke," *Gerlands Beitrage Geophysik* 11 (1912): 227–39.

13. H. O. Wood and F. Neumann, "Modified Mercalli Intensity Scale of 1931," *Bulletin of the Seismological Society of America* 21 (1931): 277–83.

14. S. Hough, *Richter's Scale: Measure of an Earthquake, Measure of a Man* (Princeton, NJ: Princeton University Press, 2007).

15. C. F. Richter, "An Instrumental Earthquake Magnitude Scale," *Bulletin of the Seismological Society of America* 25 (1935): 1–32.

16. Seismologist Perry Byerly later claimed credit for encouraging use of the phrase "Richter Scale," but I'm skeptical. Actually, the one true "Richter Scale" is a company in Pretoria, South Africa, that manufactures industrial weighing machines.

17. Richter, 13.

18. H. Kanamori, "The Energy Release in Great Earthquakes," *Journal of Geophysical Research* 82 (1977): 2981–87.
19. G. Grünthal (ed.), "European Macroseismic Scale 1998 (EMS-98)," *Cahiers du Centre Européen de Géodynamique et de Séismologie 15* (Luxembourg: Centre Européen de Géodynamique et de Séismologie, 1998).

CHAPTER 6 THE WAVE THAT SHOOK THE WORLD

1. A. Nur, *Apocalypse: Earthquakes, Archaeology and the Wrath of God* (Princeton, NJ: Princeton University Press, 2008).
2. The account of the 1929 tsunami presented here is based mostly on G. Cranford, *Not Too Long Ago . . . Seniors Tell Their Stories* (St. John's, Newfoundland: Seniors Resource Centre, 1999), and A. Ruffman, *Tsunami Runup Mapping as an Emergency Preparedness Planning Tool: The 1929 Tsunami in St. Lawrence, Newfoundland* (Halifax, Nova Scotia: Geomarine Associates, 1996).
3. Ruffman, 29.

CHAPTER 7 PREVENTION AND CURE

1. See, for instance, P. Bernard, *Annales de Calais et du Calaisis* (Saint-Omer, France: 1715), and H. D. Outreman, *Histoire de la ville et comté de Valentiennes* (Douai, France: 1639).
2. R. M. W. Musson, G. Neilson, and P. W. Burton, *Macroseismic Reports on Historical British Earthquakes XIV: 22 April 1884, Colchester.* Seismology Series, Technical Report WL/90/33 (Edinburgh: British Geological Survey, 1990).
3. R. M. W. Musson, *A Catalogue of British Earthquakes.* Seismology Series, Technical Report WL/94/04 (Edinburgh: British Geological Survey, 1994), 62.
4. *BBC News,* "Iranian Cleric Blames Quakes on Promiscuous Women," April 20, 2010, http://news.bbc.co.uk/1/hi/8631775.stm.
5. Thanks to Viviana Castelli of Istituto Nazionale di Geofisica e Vulcanologia for information about Saint Emidius.

CHAPTER 8 NEXT YEAR'S EARTHQUAKES

1. Attributed, probably wrongly, to Yogi Berra. The true originator may have been the Danish humorist Robert Storm Petersen.
2. See, for instance, S. Crampin and Y. Gao, "Earthquakes Can Be Stress-Forecast," *Geophysical Journal International* 180 (2010): 1124–27.
3. V. I. Keilis-Borok and V. G. Kossobokov, "Periods of High Probability of Occurrence of the World's Strongest Earthquakes," *Computational Seismology* 19 (1987): 45–53.
4. See, for instance, H. Becerra, "Science Is Left a Bit Rattled by the Quake that Didn't Come," *Los Angeles Times,* September 8, 2004.
5. For a full account of this story, see R. S. Olson, B. Podesta, and J. M. Nigg, *The Politics of Earthquake Prediction* (Princeton, NJ: Princeton University Press, 1989), or, for a shorter but highly readable account, S. Hough, *Predicting the Unpredictable* (Princeton, NJ: Princeton University Press, 2010).
6. W. H. Bakun and A. G. Lindh, "The Parkfield, California, Earthquake Prediction Experiment," *Science* 229 (1985): 619–24.
7. Y. Y. Kagan, "Statistical Aspects of Parkfield Earthquake Sequence and Parkfield Prediction Experiment," *Tectonophysics* 270 (1997): 207–19.

8. T. Rikitake, "Biosystem Behaviour as an Earthquake Precursor," *Tectonophysics* 51 (1978): 1–20.

9. M. Ikeya et al., "Electromagnetic Pulses Generated by Compression of Granitic Rocks and Animal Behavior," *Episodes* 23 (2000): 262–65.

10. R. A. Grant et al., "Ground Water Chemistry Changes before Major Earthquakes and Possible Effects on Animals," *International Journal of Environmental Research and Public Health* 8 (2011): 1936–56.

11. K. Wang et al., "Predicting the 1975 Haicheng Earthquake," *Bulletin of the Seismological Society of America* 96 (2006) 757–95. The Haicheng story was imperfectly known in the West until this thorough investigation was published.

12. But I'm not going to name names.

13. I have memories of long and fractious debates on this subject on Usenet in the 1990s, and I believe these still continue in various Internet forums.

14. The proceedings were published as Sir J. Lighthill, *A Critical Review of VAN* (Singapore: World Scientific, 1996). The papers in this volume provide a good overview of VAN, by both its supporters and its detractors.

CHAPTER 9 TWENTY-FIVE SECONDS FOR BUCHAREST

1. R. A. Apple, "Thomas A. Jaggar, Jr., and the Hawaiian Volcano Observatory," USGS, Hawaii Volcano Observatory, January 4, 2005, http://hvo.wr.usgs.gov/observatory/hvo_history.html.

2. F. Wenzel et al., "25 Seconds for Bucharest," in J. Zschau and A. N. Kuppers (eds.), *Early Warning Systems for Natural Disaster Reduction* (Berlin: Springer, 2003). Yes, I stole the chapter title—it was too good to pass up.

3. W. H. K. Lee and J. J. Espinosa-Aranda, "Earthquake Early-Warning Systems: Current Status and Perspectives," in *Proceedings of the Eleventh World Conference on Earthquake Engineering in Acapulco, Mexico* (Dordrecht: Elsevier Science Ltd, 1996).

4. Washington State Department of Transportation Project SR 99, http://www.wsdot.wa.gov/Projects/Viaduct/.

5. Nick Eaton, "Could Seattle Get an Earthquake Early Warning System?" *Seattle's Big Blog,* February 27, 2011, http://blog.seattlepi.com/thebigblog/2011/02/27/could-seattle-get-an-earthquake-early-warning-system/.

6. Jesse Emspak, "The Big One: Could a Warning Help?" *Discovery News,* November 27, 2011, http://news.discovery.com/tech/california-earthquake-warning-system-111027.html.

CHAPTER 10 EARTHQUAKES DON'T KILL PEOPLE, BUILDINGS DO

1. E. Gibbon, *The History of the Decline and Fall of the Roman Empire* (London: Cadell, 1837), 714–15.

2. One of the architects, Anthemius of Tralles, certainly built an earthquake simulator, powered by steam. It's not clear that he ever tested a model of the cathedral on it (Gibbon, *Decline and Fall,* ch. 40, 65).

3. A. Blakeborough, P. A. Merriman, and M. S. Williams, eds., *The Northridge, California Earthquake of 17 January 1994: A Field Report by EEFIT* (London: Earthquake Engineering Field Investigation Team, 1997).

4. N. N. Ambraseys, personal communication; N. N. Ambraseys and C. Melville, *A History of Persian Earthquakes* (Cambridge: Cambridge University Press, 1982).

5. To learn more about this effort, see NSET's website at www.nset.org.np/nset/php/english.php.

CHAPTER 11 THE PROBABILITY OF DISASTER

1. A number of pre-2010 reports, including Calais's presentation on earthquake hazard in Haiti, are available at http://web.ics.purdue.edu/~ecalais/haiti/.
2. The literature on seismic hazard tends to be highly technical and usually in obscure reports. For an easily available book-length introduction to the subject, see L. Reiter, *Earthquake Hazard Analysis* (New York: Columbia University Press, 1990). This is now a bit dated, given the speed with which the field has developed, and is not for the general reader.
3. D. Hume, *An Enquiry Concerning Human Understanding* (London: Cadell, 1772), 49.
4. B. Gutenberg and C. F. Richter, "Frequency of Earthquakes in California," *Bulletin of the Seismological Society of America* 34 (1944): 164–76; M. Ishimoto and K. Iida, "Observations of Earthquakes Registered with the Microseismograph Constructed Recently," *Bulletin of the Earthquake Research Institute* 17 (1939): 443–78.
5. M. Tantala et al., "Earthquake Risks and Mitigation in the New York, New Jersey and Connecticut Region" (report of the New York City Area Consortium for Earthquake Loss Mitigation, New York, June 30, 2003).

CHAPTER 12 STAY SAFE

1. I. Ketin, "On the Strike-Slip Movement of North Anatolia," *Bulletin of the Mineral Resources Exploration Institute of Turkey* 72 (1969): 1–28.
2. Although some have suggested a million could be killed in the next Istanbul earthquake, official projections are much lower. M. Erdik, personal communication.
3. N. N. Ambraseys and D. Jackson, "A Note on Early Earthquakes in Northern India and Southern Tibet," *Current Science* 84 (2002): 570–82.
4. D. Jackson, "The Great Western-Himalayan Earthquake of 1505: A Rupture of the Central Himalayan Gap?" in *Tibet, Past and Present*, ed. H. Blezer (Leiden, The Netherlands: Brill's Tibetan Studies Library, 2002), 147–59.
5. GeoHazards International, "Global Earthquake Safety Initiative (GESI) Pilot Project" (Nagoya: United Nations Centre for Regional Development, 2001).
6. R. Bilham, V. K. Gaur, and P. Molnar, "Himalayan Seismic Hazard," *Science* 293 (2001): 1442–44.
7. N. N. Ambraseys, "A Note on Transparency and Loss of Life Arising from Earthquakes," *Journal of Seismology and Earthquake Engineering* 12 (2010): 83–88.
8. N. N. Ambraseys and R. Bilham, "Corruption Kills," *Nature* 469 (2011): 153–55.

INDEX